DIAMONDS
IN THE
ROUGH

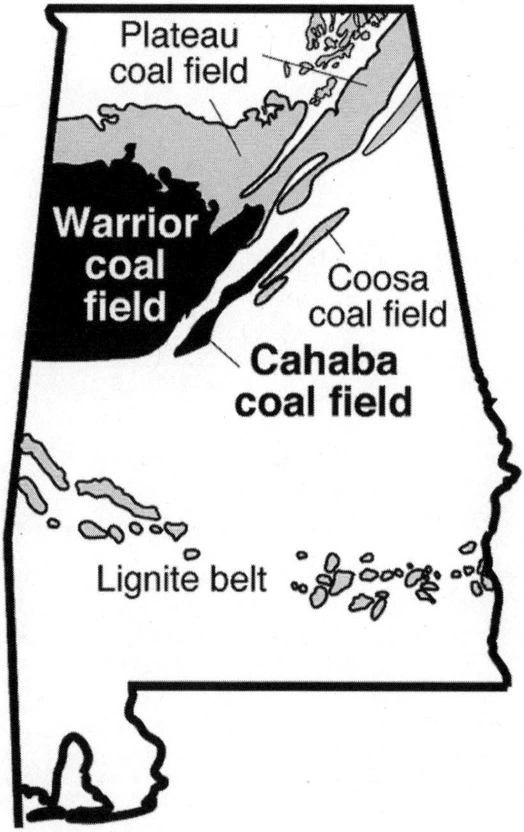

Frontispiece. Warrior, Cahaba, Coosa, and Plateau coal fields. (Courtesy of the Geological Survey of Alabama.)

DIAMONDS
IN THE
ROUGH

A History of Alabama's Cahaba Coal Field

James Sanders Day

THE UNIVERSITY OF ALABAMA PRESS
TUSCALOOSA

The University of Alabama Press
Tuscaloosa, Alabama 35487-0380
uapress.ua.edu

Hardcover edition published 2013.
Paperback edition published 2015.
eBook edition published 2013.

Typeface: Garamond

Manufactured in the United States of America
Cover photograph: South and North Alabama Railroad Bridge #72 spanning Buck Creek. Mining Engineer Joseph Squire's house is in the right foreground, town of Helena lies in the distance; courtesy of Ken Penhale and the University of Montevallo Archives
Cover design: Mary-Frances Burt

∞

The paper on which this book is printed meets the minimum requirements of American National Standard for Information Science–Permanence of Paper for Printed Library Materials, ANSI Z39.48-1984.

Paperback ISBN: 978-0-8173-5834-1

A previous edition of this book has been catalogued by the Library of Congress as follows:
Library of Congress Cataloging-in-Publication Data

Day, James Sanders, 1956–
Diamonds in the rough : a history of Alabama's Cahaba coal field / James Sanders Day.
pages cm
Includes bibliographical references and index.
ISBN 978-0-8173-1794-2 (trade cloth : alkaline paper) — ISBN 978-0-8173-8674-0 (ebook) (print) 1. Coal mines and mining—Alabama—Cahaba River Region—History—19th century. 2. Coal mines and mining—Alabama—Cahaba River Region—History—20th century. 3. Coalfields—Alabama—Cahaba River Region—History. 4. Company towns—Alabama—Cahaba River Region—History. 5. Browne, William Phineas, 1804–1869. 6. Businessmen—Alabama—Cahaba River Region—Biography. 7. Cahaba River Region (Ala.)—History, Local. 8. Cahaba River Region (Ala.)—Social conditions. 9. Cahaba River Region (Ala.)—Economic conditions. I. Title.
TN805.A6D28 2013
333.8'220976178—dc23

2012044797

For René, Abby, and Mary Afton, with much love

CONTENTS

ILLUSTRATIONS

Figures

Tables

ACKNOWLEDGMENTS

This project began as a dream approximately twenty-five years ago. Through historical study, I wanted to re-create the mining town of Piper, to put structures and faces on the rolling hills that I had walked as a child with my dad. Having met numerous "Piper People" at the annual Piper-Coleanor High School reunions, I longed to preserve their stories and to validate their memories. Earnest research was delayed a decade, but I learned in the interim that the scope needed to be larger than I first imagined. Therefore, this work represents fifteen years of diligent investigation and, I trust, constitutes a fitting testament to the miners and families of all the towns in Alabama's second largest coal field—the Cahaba.

Countless staff members at several institutions assisted with my research. First and foremost, the University of Montevallo provided time, resources, and encouragement over the years. My history colleagues—Robert Barone, Wilson Fallin, Clark Hultquist, and Ruth Truss—have supported me since my arrival at UM. Other faculty members in the Department of Behavioral and Social Sciences have expressed interest, and I gained priceless administrative support from the former chair, Susan Vaughn, and from Amanda Fox. UM funded several research and special projects grants, travel expenses, and a semester-long sabbatical in support of my research. Staff members at the university's Carmichael Library worked tirelessly, particularly with interlibrary loan, to acquire necessary (and often obscure) sources. Finally, Tiffany Roskamp-Bunt and Justin Barron contributed their technological expertise in preparing pictures and images.

Other individuals and institutions supported effective research as well—the Archives, Microfilm, and Special Collections Departments at the Birmingham Public Library; Elizabeth Wells in Special Collections at the Samford University Library; the Ralph Brown Draughon Library at Auburn University; Debbie Pendleton, Steve Murray, John Hardin, and others at the Alabama Department of Archives and History; Bobby Joe Seales at the Shelby County Archives; Alex Sartwell and Lewis Dean at the Geological Survey of Alabama Library. Colleagues on the Alabama Historic Ironworks Commission—Jim Bennett, Marty Everse, Tom Land, and Mike Mahan—gave advice and encouragement at various times. Doctoral committee members at Auburn University—W. David Lewis, Larry Gerber, Tony Carey, and David Whitten—provided sound feedback in revising the initial draft. Special thanks goes to Wayne Flynt—major professor, mentor, colleague, and friend—who read numerous drafts, offered sage counsel, and provided timely encouragement throughout the process.

Several organizations afforded opportunities to present and/or publish portions of this manuscript. "The Convict-Lease System in Alabama, 1872–1927" appeared in the *Gulf South Historical Review* (Spring 2006), and the *Alabama Review* published "Dealing in Black Diamonds: Joseph Squire and Alabama's Early Coal-Mining Operations" (January 2011). Two articles—"Coal Mining" and "Mining Labor"—are included in the *Encyclopedia of Alabama*. Presentations at the Alabama Historical Association, the Society of Alabama Archivists, the Southern Historical Association, and the Southern Industrialization Project provided venues for discussing my findings with other historians.

Numerous individuals offered their stories, time, and resources to assist with my research. Ken Penhale preserved and organized a trove of letters and documents from Joseph Squire's dilapidated barn and then willingly opened his files for my research; his assistance and friendship are priceless. Henry and Rose Emfinger worked selflessly and tirelessly to create and maintain the Aldrich Coal Mine Museum, and they are always eager to share artifacts, photographs, and information when needed. Charles Adams opened his home and provided newspaper clippings, personal accounts, and stories of Blocton, and Marshall Goggins's photographs added a human touch to the mining process. Elizabeth "Miss Lizzie" Frost and Dora Grace Smith offered an afternoon of hospitality and information about the Little Gem mines in Dogwood. Judge M. O. Cleveland welcomed me on a visit to William Phineas Browne's homesite, and Jim Lewis shared his family connections to Browne. An interview with Dr. L. C. Parnell Jr. revealed the professional and family history of Adrien Sicard, and Virgil Rice provided a windshield tour of several truck mines on a rainy day. Marlene Hunt Rikard engaged me in conversation, offered initial suggestions, and assisted with her research on the Tennessee Coal, Iron, and Railroad Company. Jack Bergstresser provided expertise in industrial archeology, mineralogy, and topography, and Everett Smith helped in understanding geological formations and the scientific aspects of mining operations. Finally, Douglas Blackmon exchanged information about the convict-lease system, and Michael Williams shared key biographical data dealing with Isaac Taylor Tichenor.

I am indebted to local historians who preserved their community histories and thereby provided "a personal touch" to the story of coal—Henry and Rose Emfinger (Aldrich); Vicky Clemmons and David Daniel (Bibb County); Charles Adams (Blocton); Ken Penhale and Martin Everse (Helena); Marie Butler (Margaret);

Eloise Meroney, Clark Hultquist, and Carey Heatherly (Montevallo); James Walker (Piper); Tommie Harrison (Wilton).

As I already mentioned, "Piper People" remain central to my interest in coal mining. Special appreciation goes to Harold, Betty, Tommy, and Bonnie Campbell; to Thomas M. and Marie Samsal Fancher; to Harry Fullman; to Walter Gardner Jr. for "By-Gone Days"; to Elizabeth Samsal Kendrick; to Howard and Angela "Doll" Milling for a map of Boothton and baseball stories; to Cecil Sewell Sr.; to Jess Shepard for meticulous drawings and detailed explanations; to Jim and Nona Terpo; and to James Walker for *The Struggle and the Joy*.

Hawkinsville—a community along the road leading to Piper—was home to the Day family. My paternal grandparents, Ezekiel Caraway and Mamie Trott Day, lived their part of this coal-mining history with six children—Marjorie, E. C. Jr., Herschel, Lorraine, Robert (Bob), and Francis. Maternal grandparents, John Alexander and Nellie Allen Sanders of Wilton, raised five children—Leonard (Bo), Mary, Nell, Johnnie, and Louise—and introduced me to the history of railroads. My parents—Herschel and Louise—and my siblings—Bud, Norfleete, and Joe—contributed to a secure and nurturing home, and they continually supported my endeavors. Additionally, Tillman, Phyllis, and Rhonda Davis welcomed me into their family and provided encouragement through the years.

Finally, the love, support, and assistance exhibited every day by my wife, René, and our daughters, Abby and Mary Afton, are beyond compare. Truly, this work represents a combined family effort, and it is to those three wonderful women that I dedicate this book.

INTRODUCTION

The Cahaba coal field stretches like a sleeping giant across central Alabama. Extending sixty-seven miles through St. Clair, Jefferson, Shelby, and Bibb Counties, the Cahaba field spawned numerous coal-mining operations during the late nineteenth century and the first half of the twentieth. Generally forgotten or ignored due to the growth of the Birmingham District and the more famous Warrior field to the north, the Cahaba field possesses a history that both coincides with and deviates from the development of its neighboring regions. Cahaba coal contains properties (e.g., sulfur content, moisture content, fixed carbon content, bulk density, and ash content) that differ from the resources of the captive mines within the Warrior field.[1] Consequently, technological considerations may differ as well. Many of the capitalists who developed the Cahaba region are identical to those of the Birmingham area: Truman H. Aldrich, Henry F. DeBardeleben, and James W. Sloss, among others. On the other hand, a plethora of small, independent enterprises introduced other entrepreneurs—some southern, some northern, some European—to the Cahaba field. Numerous communities developed around the mining operations, and these settlements generally adhered to the standard model of American coal towns. Nevertheless, each community reflected the image and character of its owner(s), its management, and its inhabitants.

Thus, a historical account of the inception, development, boom, and bust within the Cahaba coal industry is multifaceted. An in-depth study must consider geographical and geological characteristics, mining techniques and technological advancements, economic development and capitalistic ventures, and community life and social trends. But at the center of this story will be its people. Whether operator or miner, management or labor, union or nonunion, white or black, immigrant or native, everyone involved in the Cahaba coal field left a mark for posterity. Many local historians have depicted the lives of these people, but usually in episodes or segments that fracture the whole into bits and pieces. Entrepreneurs emerge, or workers and the unions they formed move into focus. But seldom do engineers and scientists, entrepreneurs and miners occupy the same canvas. Largely obscured today by pine trees and kudzu, the mining districts of the Cahaba coal field changed the lives of numerous individuals and families. This impact—this indelible and eternal influence—holds the key to unlocking the legacy of Cahaba coal.

In actuality, the history of the Cahaba coal field consists of two parts—the exploration of the field and the creation of company towns. The first phase focuses on nineteenth-century events and encompasses the discovery, exploration, and mapping of the coal lands. Challenging Gavin Wright's conclusion that the Civil War represented the watershed in southern industrialization, this examination of the Cahaba field establishes viable coal-mining operations in antebellum Alabama. Contrary to Wright's assertion that southern manufacturing efforts stagnated in the 1850s, Cahaba's story aligns more closely with J. Mills Thornton's argument that Alabama experienced an "economic miracle" in the last antebellum decade. Deliberate railroad expansion, revived banking systems, and diversified manufacturing interests combined to transform the face of economic development within the state. Still based primarily in agriculture, Alabama's economy gained momentum from capital generated by a growing demand for cotton. Coincidentally, Alabama's increasingly diversified industry expanded rapidly in areas infiltrated by railroads. According to Thornton, by 1860, Alabama had established a veritable "takeoff point" for its industrial progress.[2]

Curtis Evans further debunks Wright's thesis in his biographical study of Daniel Pratt, and the history of Cahaba coal supports his revisionist argument. As Evans reveals, Pratt worked for four decades to establish his cotton gin factory in Prattville. Employing both black and white workers and promoting railroad expansion and economic diversification, this Yankee entrepreneur established the Prattville Manufacturing Company as the largest cotton-gin maker in the United States by 1860. Certainly, he struggled with the physical and financial difficulties inherent in a frontier region, but he gained acclaim and respect as a "determined and resourceful industrialist." In sum, Evans argues that southerners did not exhibit an aversion to industrialization and that no schism existed between "reactionary planters and progressive industrialists" in postbellum Alabama. He contends further that Pratt "preached his industrial gospel" from the 1840s to the 1870s, and this time frame coincides with the developmental period of the Cahaba coal field. In fact, Daniel Pratt personified the confluence of industrial pursuits when he gained a controlling interest in the Red Mountain Iron and Coal Company in 1872.[3]

Previous historical accounts mark the beginning of coal operations in central Alabama from 1853 to 1855. This study adjusts that time frame by moving the starting date back to the late 1840s. Focusing on the endeavors of William Phineas Browne—like Pratt, a New England businessman who migrated to Alabama—and the early stages of development, Cahaba's story reveals a pioneering effort.

Fig. 1. William Phineas Brown and Margaret Stevens Browne, coal-mining pioneers of central Alabama. (Courtesy of Herbert J. "Jim" Lewis, Birmingham.)

Operating during the early phase of railroad development, Browne used an assortment of trams, wagons, railcars, and flat boats to market his coal. Transporting loads of coal to Selma, he used waterways and railways to distribute his product to customers. Traveling from Selma to Montgomery, Marion, Uniontown, and Mobile, Browne established markets throughout the southern regions of Alabama. Challenging the traditions of an agrarian, slave-based economy, he helped lay the groundwork for southern industrialization. Meanwhile, his wife, Margaret, remained at home and managed slaves, overseers, tram loads, rail car loads, and various mining and farming responsibilities. Both William Phineas and Margaret represent ambitious and resilient figures of antebellum Alabama.

English mining engineer Joseph Squire constitutes the central figure in nineteenth-century coal-mining operations. Arriving in Montevallo in 1859, he devoted a half century to exploring, testing, charting, and mapping Cahaba coal seams. Working at various times for entrepreneurs Browne, Aldrich, DeBardeleben, and Sloss, Squire accumulated a wealth of information that provides a historical record of the Cahaba, Warrior, and Coosa coal fields. Overshadowed in previous accounts by the more flamboyant capitalists, he provided the knowledge and expertise that made economic ventures profitable. Maintaining meticulous records as well as writing an autobiography, Squire provided a trove for historical research. Recording descriptions of the Cahaba coal field in the 1890 *Alabama Geological Survey,* his observations and conclusions offer interesting comparisons with twentieth-century mining pursuits. Moreover, Squire produced a map to accompany the *Survey* that attempts to depict three-dimensional detail within two-dimensional space. In addition to providing important topographical and geological information, Squire's hand-crafted imagery constitutes a work of art.[4]

Using information generated by Squire and others, Aldrich, DeBardeleben, and Sloss determined to develop and exploit the resources of the Cahaba coal field. Beginning at Aldrich and Helena, this financial triumvirate learned their trade in Cahaba before staking their claims within the Birmingham District. In partnership with his father-in-law, Daniel Pratt, DeBardeleben purchased the abandoned Oxmoor furnace in 1872 and conducted an all-important coking experiment in 1876. Determining Warrior coal to be more suitable in the coking process, these three magnates abandoned Cahaba coal for the northern region. Eventually taking separate, but related paths, each capitalist achieved success in his own right. Due to the volatile nature of the Gilded Age economy, all three experienced disappointment as well. Ultimately excluded from the Tennessee Coal, Iron, and Railroad Company (TCI), DeBardeleben and Aldrich returned to their entrepreneurial roots in Cahaba. Their pioneer efforts in establishing mining communities throughout the coal regions of central Alabama set the stage for the second phase of Cahaba's history. In contrast to the linear, chronological account of nineteenth-century developments, Cahaba's twentieth-century story prompts a thematic approach.

Ostensibly, Cahaba's unique qualities, as described by Squire, ended with the turn of the twentieth century. Even though every company town developed its own particular character, coal towns in the Cahaba field fit the model applicable throughout the Appalachian coal-mining states. In general, coal fields went through a period of small-scale proprietary development before more extensive collieries, which often included planned communities, made an appearance. Convict leasing manifested itself sparingly in Cahaba mines, gaining a foothold only at Aldrich, Belle Ellen, Helena, and Lucile. Paternalistic relationships reflected the personalities of owners and operators, and the Tennessee Coal, Iron, and Railroad Company mines at Blocton exhibited that organization's more formal system of welfare capitalism. Unionism came to the Cahaba field during President Franklin D. Roosevelt's New Deal era, and local strikes reflected state, regional, and national attempts to organize. Consequently, the twentieth-century history of Cahaba coal closely parallels that of other regions. Mining operations in Pennsylvania, Maryland, Virginia, and Colorado exhibited similar trends, but coal quality, niche markets, and colorful people exerted their influence to produce a flavor and quality of life that makes the Cahaba field distinctive.

In effect, the twentieth-century coal-mining experience differs significantly

Fig. 2. Joseph Squire, mining engineer and author of "Report on the Cahaba Coal Field" in the 1890 *Geological Survey of Alabama*. (Courtesy of Ken Penhale, Helena.)

from that of the previous century. Even though Aldrich and DeBardeleben continued to influence mining operations, a second generation of entrepreneurs sought their fortunes at numerous points across the Cahaba coal field. Opening slopes at various places along multiple seams, independent capitalists established coal-mining communities that altered the face of the Cahaba landscape. Promoting migration from rural areas, these towns constituted a rudimentary form of industrial urbanization. Incorporating European immigrants into the white and black mix of miners, Cahaba coal towns also produced a cultural blend that defined the surrounding area. Convict leasing, welfare capitalism, and unionization constitute overlays that affected the social fabric. Integrated into the coal-mining experience, these influences created a multifaceted culture that generated a common bond among the people of Cahaba.

Relatively short-lived, the Cahaba coal-mining experience effectively ended in the mid-1950s. Mechanization and strip mining allowed some enterprises to continue for a time, but larger economic and technological trends forced the industry's decline within the Cahaba field. Still, for a little more than a century, mining operations dominated the Cahaba coal regions of central Alabama. Combining extant written sources with oral accounts and personal recollections, this study attempts to reconstruct the historical moment that defined the Cahaba coal field.

1
DISCOVERING AND MARKETING COAL

1815–1859

The discovery of coal along Alabama's Cahaba River is legendary. Named by Professor Michael Tuomey, the first state geologist, the Cahaba coal field includes the site of the first systematic extraction of coal in Alabama. However, initial discoveries may be traced back to 1815 when several veterans of the Battle of New Orleans made their way across the Mississippi Territory and into the area that would become central Alabama.[1]

These early settlers, led by Major Jonathan Mahan, numbered eighteen when they arrived at the head of the Little Cahaba River. Finding an Indian camp situated at this confluence of present-day Mayberry, Mahan, and Shoal Creeks, two-thirds of the party settled down with Indian wives. Six others continued northward, but later returned to create the settlement of Brierfield, where they staked their claims along the waterway they dubbed Mahan Creek. According to Ethel Armes, historian of Alabama's coal and iron industries, the Mahans "built the first flat and keel boats ever floated on [the] Cahaba River to carry coal." She also gives credit to the Mahans by contending that the discovery and extraction of coal in the Cahaba field "is directly traceable to this little group of pioneer settlers and those following them."[2]

Another personal account related by Armes gives credit for the initial discovery of coal to two young boys on a hunting expedition. According to Mrs. Frank Fitch, her father, Jonathan Newton Smith, and Pleasant Fancher set up camp near the Big Cahaba River sometime in the early 1820s. Using stones and logs for their campfire, they cooked supper and then drifted off to sleep. At some point during the night, Smith awakened to find the "stones" on fire. Scared out of their wits, the boys raced home in the dark, but they later realized that they must have used lumps of coal inadvertently. At any rate, the stream that flowed into Dailey Creek became known as Coal Branch.[3]

These traditional versions complement Truman Aldrich's more historical account that claims that numerous citizens of Bibb and Shelby Counties had collected coal from the region as early as 1836. Gathered along exposed drifts or from open pits, small amounts of coal—known in the vernacular as "cornfield diggings"—provided heat for family dwellings and fuel for blacksmith shops. In fact, the Fancher Pit and Wood's Pit bore the names of families who settled near the coal outcroppings. Aldrich also noted efforts by D. H. Carter, a resident of Montevallo in 1852, who mined some coal from the Lemley Seam, hauled his loads to the Alabama & Tennessee Rivers (A&TR) Railroad, and shipped the coal to Montgomery where it was sold to blacksmiths for $6 per ton. One year later, a group of Montgomery citizens formed a company and attempted to float barges of coal down the Cahaba River only to have all but one destroyed on the shoals at Centreville.[4]

Shortly thereafter, Colonel Daniel E. Watrous, president of the newly formed Alabama Coal Mining Company (ACMC), commissioned a study by State Geologist Tuomey to evaluate 4,800 acres of coal lands controlled by the company. Having established a creditable reputation as state geologist of South Carolina, Tuomey came to Alabama in 1847 at the invitation of the faculty at the University of Alabama. He began work in May as the chair of geology, mineralogy, and agricultural chemistry, and he organized the first systematic geological survey of Alabama two months later. In January 1848, the general assembly adopted a resolution appointing Tuomey as state geologist and commissioning the *Geological Survey of Alabama.*[5]

Submitted in October 1855, Tuomey's report confirmed speculations that the Cahaba region contained vast amounts of accessible and valuable coal. Tuomey's letter also identified the more prominent coal beds within the company's holdings. The Watrous Bed consisted of a five-foot seam that outcropped along the southern boundary of the company's lands. The Pushmattahaw Beds contained five or six seams, but Tuomey recommended mining the two most prominent ones, namely a seam containing thirty inches of coal and the other with a thickness of 4.5 feet. Due to the length of these beds (approximately 1.5 miles), he estimated the available coal at just less than one million tons. The Level Beds consisted of the aforementioned Fancher and Wood's Pits where numerous coal seams measured between two and three feet in thickness. Farther to the north, the Tustinuggee Beds contained numerous seams, the three most promising consisting of coal measures of 7 feet, 17 inches, and 4.5 feet, respectively. At this point, Tuomey identified a characteristic of the Cahaba field that would consistently challenge everyone who attempted to extract coal: "As the field has never been proved by boring, and the whole being one unbroken forest, it becomes very difficult, especially where rocks are undulating or slightly inclined, to determine the number of beds superimposed upon each other."[6]

Nevertheless, Tuomey estimated that the total amount of available coal in the three major beds exceeded eight million tons. Recognizing the railroad as the only reliable and reasonably priced means of transportation, he predicted significant savings as soon as the company completed a number of branch lines, thereby connecting the mines with the main railroad. Further advocating the use of Cahaba coal in the manufacture of gas, he named Selma, Mobile, and New Orleans as potential markets. In conclusion, Tuomey expressed his appreciation "in viewing this first, really business like attempt, to open and unfold the riches of one of our great mineral deposits."[7]

Basing his historical summary on Tuomey's report, Aldrich asserted that "the first systematic attempt at the mining and shipping of coal, was made [by the ACMC] in the Cahaba coal field near its southwestern extremity, above Pratt's Ferry and on the right bank of the Cahaba river." Ethel Armes confirmed this account when she recorded that "the first regular systematic underground mining in the state [of Alabama] had been done in the Cahaba field in 1856, at a point in Shelby County, one mile west of . . . Montevallo." Later in his account, Aldrich identified William Phineas Browne of Montevallo as "the owner of a considerable tract of land adjoining the lands of the Old Alabama Mining Company." Aldrich reported accurately that Browne opened several mining pits along the same Montevallo vein tapped by the ACMC, but he established the period from 1856 to 1863 as the time frame for Browne's endeavors. On the contrary, Browne's records indicate that he moved to the Montevallo area in 1847 and began mining coal as early as 1849. Moreover, Armes counted two hundred men involved in the coal trade in Shelby County in 1850.[8]

Developing transportation networks and promoting coal sales throughout central and south Alabama, Browne pioneered the development of coal mining in the Cahaba region. His efforts followed in the wake of other extraction enterprises along the Appalachian chain. Southern commercial coal mining had begun in the Richmond Basin in the 1760s, and Virginia and Pennsylvania dominated the industry in the early nineteenth century. Coal mining commenced in Kentucky during the 1830s, and Tennessee and Alabama followed suit by mid-century. As Sean Patrick Adams opines, "there is a high correlation between industrial development and the presence of coal mining." By 1853, the year in which the ACMC first organized, Browne had established a business partnership and was well on his way to extracting, transporting, and selling coal. Certainly, the ACMC would constitute his primary competition, but credit for the first systematic mining in the state of Alabama must go to Browne, a native of Vermont who came to Alabama in the 1830s.[9]

Born in Waltham, Vermont, on July 9, 1804, the son of Phineas Browne and his second wife, Elizabeth Backus, William Phineas left home at age seventeen to seek his fortune. He worked for one year as a business clerk in Manhattan before joining his cousin, John B. Ives, on a canal construction project near Lebanon, Pennsylvania. Following this work on the Union Canal, Browne returned to Vermont to study law, passing the bar in 1829. After practicing law for two years, he rejoined Ives on another canal project near Muscle Shoals, Alabama. As Browne biographer Virginia Estella Knapp records, William Phineas embarked on "a period of business enterprises which were of several kinds—canal construction, mail steamer service, mercantile business, and real estate."[10]

From 1831 to 1849, Browne experienced many challenges, such as cash flow and labor shortages, that prepared him for a career in the coal business. After three years of canal construction in north Alabama and New Orleans, he joined a steamship company operating out of Mobile. However, numerous accidents, explosions, and other mishaps crippled the small fleet of steam ships, and the partners sold out in December 1835. Browne invested his proceeds in a general merchandise store in Liverpool, Mississippi, a small town on the Yazoo River. Marketing cotton and selling supplies constituted the primary money-making activities, but court actions forced Browne to sell at a loss early in 1836. In addition to his retail pursuits, Browne invested in Mobile real estate, speculating in properties, bank bills, and credit over the next four years. Knapp quotes Browne: "In January, 1837, I was worth by every estimate that could be made based upon the current value of my property over $200,000." Hard economic times descended in 1837, however, and Browne struggled to minimize his losses. He remained solvent for a time, but protracted litigation cost him his fortune.[11]

After falling on hard times, Browne returned to the legal profession and opened a small corn mill in the Mobile area. He also became involved in local politics, serving as a Mobile alderman in the early 1840s. Renouncing his affiliation with the Whig Party in 1843, Browne campaigned and won a seat in the legislature as a Democrat in 1845. Serving but one term as the representative from Mobile, he continued to dabble in numerous pursuits. Selling his corn-milling business in 1846, he joined the Alabama Volunteers at the outbreak of the Mexican War. Two years later, he began construction of a small hotel at Point Clear, Alabama, promoting the sea air, swimming, and fishing of Mobile Bay as the primary attractions.[12]

In February 1849, Browne sold his hotel interest and moved from Mobile to Montevallo. This move from the Gulf Coast to central Alabama developed as a result of Browne's marriage to Margaret Elizabeth Warwick Stevens on August 9, 1846. Born in Bibb County on April 7, 1824, Margaret was the daughter of Henry W. and Sarah Sterrett Stevens. Twenty years junior to William Phineas, aged forty-two, she had often remarked that she would wait to marry until she met her ideal mate—"a middle-aged bachelor having a bank account of at least $15,000." After a honeymoon trip to Boston, Massachusetts, and Vergennes, Vermont, William Phineas brought Margaret back to her home near Montevallo while he pursued his business ventures in Mobile. As Knapp reports, "Browne could not get settled during the years from 1847 to 1849. He and Margaret had no permanent home, and he had no particular business connections. They alternated between living with Margaret's people and in his Mobile lodgings. Browne had land in Shelby County west of Montevallo, but nothing of interest held him there."[13]

In the meantime, William Phineas and Margaret had their first child, Claudia. Still, Browne could not seem to focus on one particular pursuit to provide for his family. Trying unsuccessfully to coax Margaret to move to Mobile, he eventually decided to forsake the Point Clear project when she refused to leave Montevallo. As Knapp points out, "although Browne had been married for three years and had the added responsibility of a child, he still remained without a permanent home and without a definite business. It is not to be assumed, however, that he was in strained circumstances." She quotes Browne: "My circumstances for eight or ten years have been gradually improving . . . and if my property were disposed of at what I suppose it worth and invested profitably I should be in what is called comfortable circumstances nothing more than $30 to $50,000."[14]

Browne's marriage to Margaret proved fortuitous even though she often appeared frustrated and disappointed. Perhaps, instead of levying criticism, her letters served to inspire him. In January 1849, one month before his departure from Mobile, she wrote: "I feel much concerned, and sometimes distressed, about our situation. We have been married nearly three years, and your pecuniary affairs have grown worse all the time, but you must not give up and despair, my dear Willie, only . . . struggle the harder, and I think surely things will change for the better, and all will come right after a while. I fancy in my many imaginations, that at some future time we will be settled, and have a home of our own."[15]

It may be that Margaret was ready to move away from maternal control. Her mother, known as "Aunt Sallie," gained a reputation as a formidable personality in her own right. When her first husband, Robert Sterrett, died, Sallie moved to Shelby County near Montevallo. Marrying four husbands—Sterrett, Henry Stevens (Margaret's father), John Allen, and Richard Wood—Sallie managed her own affairs. One contemporary account by a traveling preacher states that he became "acquainted with her in 1836, when I first went to the university [of Alabama in Tuscaloosa], I was admirably entertained when stopping at her house." Apparently, Aunt Sallie's cooling house was a popular stopping point along the road.[16]

Meanwhile, other interested parties began to work their way into the mineral-rich regions of central Alabama. For example, Browne's father-in-law, Richard Wood, reported that a coal digger employed by a Montgomery-based company informed him that he was hired to explore the Cahaba River Basin looking for evidence of coal and iron. Also, Browne received the first in a series of letters from Philip J. Weaver, a Selma merchant and future treasurer of the Selma Gas Company, who expressed interest in developing the coal fields.[17]

Transportation remained a limiting factor, however. John Strong Storrs, who would become president of Browne's foremost competitor, the Alabama Coal Mining Company, wrote that an 1850 railroad bill had left Montevallo out entirely. He stated that "the people [are] not being willing to go 10 miles out of the way at an increased expenditure of 100 thousand dollars for one accommodation." By mid-1850, Weaver sounded more optimistic. He predicted that increased interest in the railroad enterprise would relate directly to a surge of emphasis in the coal lands around Montevallo. Consequently, Weaver and Browne formed a partnership and arranged to purchase many of the more promising areas west of Montevallo.[18]

About one year later, Browne began to correspond with George Oscar Baker. Born in Philadelphia in 1828 and trained as a mechanical and mining engineer, Baker would move to Selma, Alabama, in 1855. Apparently, he had designs as early as 1851 on returning to Alabama to seek his fortune in the coal business. He cautioned Browne to "be very careful not to compromise me with Weaver. I may need his favor should I return to Selma. I would like to have you destroy my letters." At this juncture, Browne left Alabama for an extended trip (approximately two years) to New England. He returned to Alabama in November 1853, prompted by letters from his sister-in-law. Browne learned that new stores and hotels added to Montevallo's growth almost daily. She wrote Margaret that "if Brother William has any thought of ever doing anything with his coal mines now is the time." Baker remained involved, proposing that he contact Weaver and offer $10,000 for his portion of the lands held jointly with Browne. Again, Baker cautioned Browne: "Just keep quiet as possible & I think it will come out all right as now fixed. Don't open any new places for him to see. Nobody sees your letters but myself."[19]

On the other hand, Weaver informed Browne that he had gained no substantive information from any of his contacts within the coal trade. Therefore, he advocated a deliberate and conservative approach to mining operations. Weaver remained guardedly optimistic as the partnership developed in the later months of 1853. He and Browne discussed the expense of labor, free miners as well as slaves. Weaver stated his belief that "half a dozen miners will make it rather more expensive than is necessary to develop the value & extent of our possessions." But acknowledging Browne's expertise in this area, he resolved to "leave this for your determination."[20]

A letter from R. H. Jackson of Greensboro, Alabama, reveals the entrepreneurs' solution to their labor needs. Having received Weaver's request on Browne's behalf to hire a number of his slaves, Jackson replied: "I regret very much that I did not get his letter sooner, as I have hired out several of the Negroes for the next year. However you can hire the balance of them if you wish at the same rate that I hired them at [*sic*], which is $225 for the men, and $125 for the women, and the same for the Boy Lewis, you paying all expenses. If you conclude to hire them please let me know immediately." Slave labor was not the only option open to Browne. One free laborer presented the following letter of introduction from Thomas B. Scott: "The bearer of this is a sober and industrious man, a very brisk worker, and a pretty fair mechanic and from my idea of your wants, most admirably adapted. I should be very glad you would give him a trial, as I am persuaded you would consider him an acquisition."[21]

Thus, Browne and Weaver commenced their joint coal-mining enterprise early in 1854. Relying on slave labor and rudimentary transportation links, they experienced many of the challenges that had plagued miners in Virginia's Richmond Basin for decades. As Adams indicates, problems abounded when "engaging in an industrial pursuit while situated in a region without an economic infrastructure geared toward expansion." In fact, the use of slaves kept labor costs high and inhibited the introduction of innovative mining technology. This dependence on slave labor to the detriment of technological improvements compromised efficiency, productivity, motivation, and discipline and exacerbated production delays, unanticipated expenses, and other setbacks.[22]

In February, Weaver informed Browne that "there is not a peck of coal in town [Selma]. Can't you send some down directly; it would find a ready market." Horace Ware, future president of the Shelby Iron Works, contacted Browne in midsummer, stating that he "would like very much [if] you would let me have 2 or 3 loads of dry coal with as little as possible of the very fine coal. The wet & fine coal will not coke well & the wet is very heavy to haul." Later in the year, Weaver requested a five-ton carload—"I should like to have you send no lumps as much as possible as I desire it for my own use to burn in stoves." Therefore, 1854 represented a year of promise. One of Browne's relatives who lived in Boston offered the following encouragement in response to Browne's report of the year's activities: "I am glad that your coal mine is promising & hope it will equal your most sanguine hopes. I send by mail a paper containing some remarks on the coal trade which may interest you. The prices at Philadelphia are double what they used to be & if coal keeps up for a few years, you will make a fortune."[23]

Year's end held fewer prospects for success. Low demand, labor problems, and transportation challenges combined to thwart Browne's enterprise. James Adams, Browne's agent in Selma, notified him in early December that he had sold no coal thus far. In spite of poor grammar and spelling, Adams's frustration and bitterness were clear: "I have got anuff of cole hear to do some time I think and so wold you if you was to come hear and stay as long as I have and do nothing and pay board you had better pick up your cole at the station on the plat forme and wate until the RRRoad get better. They have not done eney thing to the end of the track as they promised nor I don't think they will for some time."[24]

Apparently, poor markets persisted in south Alabama throughout the winter, thus prompting Browne to levy some harsh criticism toward Adams. The latter responded in kind: "You seem to think that I am trying to swindle you. I don't want you to even hint that to me again. . . . I hope you will consider that I am trying to do the best I can."[25]

In addition to fluctuating demand, Browne also faced the challenges of consistent supply. Father-in-law Richard Wood reported in late December that he and two slaves—Jack and Abram—had brought three loaded cars to the depot the previous day. The slaves would return to the mining camp to load an additional three cars, and Wood pledged to continue loading three cars per day as long as the coal supply held up. As a contingency, he recommended that they load one car with wood in the event that the railroad company charged a flat freight rate whether or not the cars were loaded.[26]

Building a reliable labor force generated a host of concerns as well. D. W. Sterrett of Camden responded to Browne's request for financial assistance by agreeing to a joint purchase of ten slaves. He insisted that he could not finance that many himself, but suggested that Browne borrow some money from Weaver "with good security." Browne's wife, Margaret, sent him word in Mobile of an upcoming slave auction: "Papa [Richard Wood, her step-father] says there are some thirty or forty Negroes to be sold at Plantersville, on Monday. You can remain and attend the sail [*sic*] if you wish, with the promise that you purchase some. . . . Jack says they are getting on pretty well at the pit, he says they need another hand. . . . Pa has been to the pit several times to see how they are getting on."[27]

Transportation assets also presented a continual drain for Browne. At one point, officials of the A&TR Railroad informed Browne that he need not attempt to ship more coal until freight charges could be paid. When markets declined, James Adams suggested purchasing a barge for transporting coal and lumber along the Alabama River. In addition, Margaret reported that the wagoners were idle. No railroad cars arrived during the week, and there was no place to put the extracted coal. The teamsters were growing impatient.[28]

As if the start-up problems were not enough, Browne's challenges compounded in 1855. Having addressed the issues of supply and demand, labor, and transportation, he soon faced the two-headed specter of quality control and competition. Mobile agent J. Cartwright reported to Browne in the spring of 1855: "Blacksmiths do not like it [coal] on account of too much dirt. They say it chokes up their fires. . . . The prospects for selling much through the summer is [*sic*] not very promising. I have collected none of the accounts (you mention [in letter of 29 April]) yet, but will make an effort to do so in a short time. . . . I think your coal would suit the blacksmiths if it could be got entirely free from dirt."[29]

Price, instead of quality, was the critical issue in Selma. New agent R. H. Kerr reported the loss of a sale because he could not sell Browne's coal at $4.50 per ton.

He surmised that the buyer negotiated that price with a company out of Montgomery. The combination of quality, price, and competition prompted John H. Murphy & Company of Montgomery to discourage Browne from sending more coal to that city. Issuing an optimistic challenge, however, an official stated that "when you can ship at a reasonable freight, good grade coal, they will sell here during the winter—also good Blacksmith coal."[30]

The situation in Montgomery epitomized the development of competition within the coal market. Kerr expounded on the issue as it related to his lost sale and offered the following conjecture to Browne: "I have for some time been of the opinion that the Co. would take the advantage if they could; I hope we will be able to repay them. I don't wish to do anything dishonorable, but at the same time I hate to be out done. . . . I am fully of the opinion that the Co. intends putting the price of coal down this winter provided they will be able to keep going. . . . If I was able I would make them suffer for such conduct, if they are guilty of under bidding in price." Kerr's reference to "the Co." introduces the commencement of mining operations by the ACMC. This enterprise, established near Montevallo, would prove to be Browne's nemesis as well as his source of motivation in developing his coal interest. In addition to competing for transportation assets and coal markets, the ACMC literally competed for coal. While Browne's cutters worked in their pits, the "Company" opened mines of its own—all of which tapped the same coal seam, three miles west of Montevallo.[31]

The remainder of 1855 marked Browne's attempts to establish his business as a solvent entity. Understandably, sales remained slow throughout the summer months, but his agents worked continually to garner back payments. W. R. Bill of Selma vainly attempted to collect from purchasers. He informed Browne: "I shall go around and call on them once more and then sue on all that are not paid. Thoug[h] perhaps it would not be policy to sue some of them for nothing could be mad[e] by it they having no property that could be got hold of." He reported further that a Mr. Hagy had been unable to fill Browne's order for coal barrels due to a fire at his shop. Now reestablished, Hagy could begin work, but "he had done nothing to your cask since my last and he could not tell when he would be able to do anything that required an outlay of money." Other correspondence dealt with outstanding balances on cotton shipments, notices from attorneys and collection agencies, searches for missing oxen, and stock purchases from the A&TR Railroad Company.[32]

Orders for coal began to trickle in during the fall. Robert Smith of Selma contacted Browne stating, "I want you to send me down a cart load as soon as you get this letter. . . . I have bin [*sic*] buying charcole [*sic*] and can't get anymore. I want the cole [*sic*] right away. Don't disappoint me." In addition, partner P. J. Weaver informed Browne: "I am in want of five tons of good lump coal for my own use. Will you please send down that am[oun]t for me."[33]

Browne closed out the year by taking positive steps to bolster his enterprise. He requested the emplacement of a switch to facilitate a tramway for transporting coal from the mines to the railroad, and he traded $2,500 in stock from the South Alabama Mining, Manufacturing, and Transportation Company for additional coal lands. Before year's end, Weaver informed Browne that no action had been taken concerning the rail switch, but a short message from George O. Baker set the tone for a positive turn of events in 1856. Baker simply informed Browne: "Will send up a/c [account] of coal [as] soon as possible." With Baker in charge of affairs in Selma, Browne had an ombudsman who would exercise initiative and devote the personal attention required to build this coal enterprise.[34]

However, even with Baker's efforts, Browne continued to experience numerous difficulties. Unrefined mining techniques and interruptions in transportation hampered his ability to satisfy a growing demand for coal. Increased quantities often resulted in poorer quality, thus leading to customer dissatisfaction. Also, weather conditions frequently affected demand and thereby thwarted adjustments to supply. Moreover, Browne could not be at every critical location. When he remained in Montevallo to oversee production at the mines, he relied on Baker to manage sales at Selma. While he was away from home developing new markets in Montgomery, Mobile, Marion, and Uniontown, his wife, Margaret, was forced to supervise slave labor, monitor mine operations, and manage tramway and railroad procedures. Thus, Browne struggled continually to develop an efficient system that effectively integrated the various steps of coal production from mine to market.

In January 1856, Browne's rudimentary mining techniques could not keep pace with the demand created by an unusually cold winter. Retail agent J. Cartwright of Mobile informed him that cold weather resulted in the depletion of both wood and coal stores. A citywide shortage had persisted since early December, but he had no lump coal to satisfy consumer heating needs. Baker recounted similar conditions in Selma: "No coal at either coal yard. [Three] in[ches] snow & snowing awful times with no coal."[35]

Furthermore, as Browne struggled to bring supply in line with demand, coal quality became an issue. Baker complained about having to rescreen more than forty tons of coal to remove the dirt. Cold, wet weather complicated this procedure and left much of the screened coal in poor condition. Fortunately, Baker salvaged enough good coal to fill an order from P. J. Weaver, but he requested that Browne make allowances for the poor quality and extra work involved. Similar reports arrived from Mobile where agent Cartwright bemoaned the poor quality of the coal. He reported selling an occasional barrel of fine coal and told of one man who agreed to purchase three or four tons at $4 per ton. Nevertheless, he advised Browne that "it may probably be a good time to sell real estate in the latter part of the spring or beginning of the summer months." Hence, Browne's prospects for success in his newly established coal business remained grim in mid-1856.[36]

As Browne worked feverishly to market the coal, Margaret tended to other affairs around the farm. Her letters reveal the remarkable details of an antebellum woman's joint enterprise with her husband. She leased two houses and a garden for $5 per month, rented two cabins to Negro inhabitants for half that rate, converted the old carriage house to an abode, and rented their final holdings—one cabin in the vicinity of the old coal pit and another near the new pit—by mid-June. In the meantime, she processed a shipment of meat forwarded by Browne, noting that she received fourteen shoulders in all. She informed Browne: "I had the meat examined. It is not spoiled, I think, but is full of skippers [moth-like butterflies]. I will have it picked and washed this morning." She also informed her husband that she had three servants working plows and that the garden was looking good following a recent rain. She had gathered several messes of squash, some green corn, and numerous cucumbers and had made blackberry jam, but she was not optimistic about the sweet potato crop. Margaret acquired a mule for plowing, but complained about their servant, Sam, who seemed to be, or pretended to be sick at least half the time. All of this responsibility prompted her to remind her husband not to "forget your humble home or its *inmates* [emphasis added] for my thoughts and dreams are all of you."[37]

As the winter of 1856–1857 approached, business picked up. In fact, this season represented the first time that Browne's supply of coal came close to meeting all demands. Furthermore, competition with the ACMC increased as both enterprises vied for markets. Agent Cartwright of Mobile ordered two casks of lump coal for his own use in mid-October. Still, he remained uncertain concerning the supply of coal for retail during the winter months. In a postscript to his order for coal, he asked Browne: "Please write me on receipt of this whether there is likely to be any supply of Shelby coal for Mobile this season."[38]

George Baker appeared more optimistic. He believed that if Browne could secure five teams of horses, then he could sell all the coal delivered to Selma. He related the following prognosis to Browne: "I find that the Coal Co[mpany] will almost certainly be out of coal. Please leave no stone unturned to secure [horse] teams. I have arranged with the central warehouse to sell all our coal in Hhds [hogsheads]. They sold 100 [barrels] for the Coal Co. last mo[nth]. . . . Every effort must be made to get coal forward. I can no doubt sell 15 tons per day at the start, & ought to have 100 tons on hand to get up a sort of confidence among the people as to our ability for keeping up a supply." Several days later, Baker informed Browne that he had "so many applications for coal [that] I feel half afraid to go in the street."[39]

Margaret also remained involved in this process. While Baker worked to obtain rail cars and transport them to Montevallo and Browne canvassed Selma, Montgomery, and Mobile for customers, Margaret managed affairs at home and at the mines. At some point in this mad rush to gain, fill, and transport carloads of coal, Margaret hastily penned a note to William Phineas notifying him that he could expect two carloads of coal in the immediate future.[40]

By the first of the year (1857), prospects for success looked more promising. On the day after New Year's, Baker informed Browne that he could not compile an account for the initial coal shipment. His freight bill indicated a total weight of 29.05 tons, but he needed the number of each car and the name of the hauler to maintain accurate books. He also indicated that, with the onset of cold January weather, competition had increased significantly. He wrote Browne: "The Ala[bama] C[oal] M[ining] Co[mpany] are running it on to me pretty hard. They keep their carts going about the streets, offering their coal for sale, & I expect will head me off considerable [*sic*] at the start." On the other hand, the ACMC had resorted to double-screening to enhance the look of its coal, but the process would necessarily cause its prices to skyrocket.[41]

A few days later, Baker cautioned Browne not to let him run low on coal. He reported daily sales averaging five tons and predicted a continual improvement. He hoped to increase sales to ten or twelve tons per day by the end of the month. Continuing in this optimistic tone, he informed Browne that the ACMC was "fizzling out." Baker continued: "They say, 'we are only bringing down this good coal for a bait & will slack off.' We will show them a thing or two before many days. I will have all their customers before spring. On Saturday I sold a ton to one of their stockholders."[42]

The next several days did not bode well for the entrepreneurs, however. Baker urged Browne to drop his charges for extraction by 25 cents to keep the retail price of coal in line with that of the ACMC. Other factors began to squeeze Baker and Browne as well. City officials in Selma threatened to curtail Browne's coal shipments unless Baker made some payments on outstanding debts. The ACMC also courted some of Browne's teamsters to gain control of the oxen and horse teams. In sum, Baker fought a cash flow problem in mid-January 1857.[43]

Still, Baker experienced a "good" problem, of sorts. Orders continued pouring in. On the same day that he bemoaned poor cash flow, he informed Browne that he had only three tons of coal remaining on his platform. He stated further, "We have delivered 30 loads today. It is awful cold & I have had no time but to attend to delivering coal." Several days later, he notified Browne of his successful negotiations with Horace Ware, president of the Shelby Iron Works at Columbiana. According to Baker, Ware wanted a guaranteed supply of fine and coarse coal. In the same letter, Baker added that he had received an order for one hundred casks to be delivered to New Orleans. "Don't you wish we could fill them all! My mouth waters to do it," he crowed.[44]

By early February, supply had declined significantly, and Baker was considerably more realistic. He relayed word that the ACMC had extracted one hundred tons of coal in one day. Obviously that news boosted the morale of the company agents, and the discovery of five hundred tons in the company's Irish Pit promised good fortune for their foreseeable future. Baker sounded despondent as he informed Browne that neither he nor the ACMC had any coal in Selma at that time. He stated: "I presume it can't be helped, but when it comes rolling down in plenty the winter will be over & no sale for it this side of Mobile."[45]

Even with this more pessimistic outlook, Baker believed the coal business had a promising future. In reality, he was too busy to remain pessimistic for long. Immersed in all aspects of the coal trade, he responded adroitly to numerous demands and opened new markets for their business. After expressing thanks for a recent shipment of coal, he informed Browne that the first load would go to Marion the following morning, thereby extending their markets west of Selma. He noted that "The Ala Co. are here & talk very friendly with me & act so, but they can make nothing by duplicity." Baker reported that the company had no coal stockpiled, and he surmised that officials had been forced to convert some stocks into working capital. He, too, was low on coal, but he expressed to Browne: "We presume you will soon arrange to connect with the coal road, or make a short haul, so as to keep us in coal with a few teams." Baker's presumption was unfounded. Apparently, Browne continued to experience transportation problems. Even though he and Baker had put tremendous efforts into building a tram connection from the mines to the main railroad line, this arrangement remained incomplete. The situation prompted Baker to plead with Browne: "Can't you hire a team at some price to keep me in coal until the branch [rail]road arrangement is certain? We have now been out of coal since Monday last & . . . our best customers are leaving us."[46]

Bemoaning continual fluctuations in their ability to supply outlying areas such as Cahaba and Marion, Baker encouraged Browne to take immediate and drastic steps to ensure a continuous supply of coal. Although completion of the branch rail line seemed imminent, Baker insisted that Browne not rely solely on that means for transporting coal. Obviously, Baker felt the pressures of increasing demand that Browne could not sense. He informed Browne that one man from Montgomery stepped off a river steamer and ordered ten casks and that orders amounting to thirty tons of coal remained outstanding in his books. According to Baker, they must either stop taking new orders or increase their supply to meet rising demand. In a final note, Baker reminded Browne: "You have seen enough of the railroad to know that there is no dependence to be placed in it & we must keep on hauling to the station until we are thoroughly underway on the Branch road."[47]

As demand for fine coal declined in late February, Baker instructed Browne to send nothing but lump coal. Informing Browne that the ACMC had decided to reduce its price per ton to sell "dirty" coal directly from the mines, Baker encouraged Browne to increase his supply of good, "clean" lump coal. He told Browne that "it will certainly not do to be without coal now when the purchase or hire of two mules only stand in the way. We can sell 500 tons this month which would put over $1300 in your pocket & the purchase of two mules are a mere trifle in the way." Furthermore, expressing a sense of urgency and an air of frustration, Baker stated: "We have now spent two months of the season waiting for the railroad to be finished & my time is too important to me to be sitting here waiting for something to turn up. Just send on the coal & you will get money fast enough to put you through."[48]

In spite of Baker's pessimistic entreaties, Browne seemed to maintain a more positive outlook. A note from his long-time friend, Alanson Saltmarsh of Cahaba, congratulated Browne on his success. Perhaps Browne enjoyed a more detached perspective that allowed him to remain optimistic about the prospects for success, whereas Baker suffered under the weight of daily challenges of balancing supply and demand. Nevertheless, Baker acknowledged Browne's increasing assets when he discussed sharing the business with potential investors. He estimated Browne's interest in the coal-mining business at $80,000, about $10,000 less than Browne's own estimate. At the same time, he informed Browne of daily sales totaling thirteen tons and commented that "if this would keep up for 12 mo[nths], something might be made to pay up for losses in December, Jan[uar]y & Feb[ruar]y." In mid-March 1857, Baker sent the following assessment to Browne: "I fear we are still ahead of our time, & must wait until some plan is devised for getting the coal down at less expense." Thus, it seems that Baker struggled continually between a realistic approach that called for caution and a more positive outlook that promoted risk-taking.[49]

This on-again, off-again approach to the coal business—typical of nineteenth-century enterprises—may be illustrated by several events in the spring of 1857. On one hand, Baker continued to deal with the labor problem. In March, he offered to sell a twenty-five-year-old male slave to Browne for $1,300. Assuring Browne that he was getting a bargain in light of the current slave market, Baker vouched for the slave's physical health and dedication to work. On the other hand, Baker seemed fixated on the perennial cash flow problem when he advised Browne to forsake his plans to expand his trade along the Alabama River. He explained projected expenses in excruciating detail and saw no way to make money on the deal.[50]

Still, Baker continued to make transactions within the coal trade. Two months after his discouraging letter to Browne, Baker wrote that he wished to receive twenty-five to thirty tons of coal per week throughout the summer. He told Browne of his plan to stockpile 150 tons as security for his regular customers, and he forwarded orders for ten carloads of coal already on the books. Furthermore, he informed Browne that he had negotiated these sales at the rate of $6.50 per ton and that he hoped to receive an additional order for one hundred tons to be delivered to Uniontown. In his usual realistic brand of optimism, he encouraged Browne to push coal forward in case winter weather interrupted transportation assets. But, lest he become too optimistic, Baker also informed Browne: "I shall hold on for you as long as I can. I am working hard to get some monied parties to take hold with us, but . . . I see no chance for me to get money enough. I am getting in debt here: slowly."[51]

Nevertheless, the summer seemed to bode well for Baker, and coal markets expanded in the fall. In early October, he reported that he had fallen behind in deliveries to his Marion and Burnsville customers. He needed at least fifty tons to overcome the discrepancy, and prospects continued to improve as cooler weather set in. By late October, he was nervous again. Rumors abounded concerning the suspension of operations by the ACMC, and Baker feared that he would run out of coal.[52]

The rumors were unfounded, but new opportunities did develop for Browne and Baker. Orders came in from Uniontown, and Frank Steward of Mobile contacted Browne asking to retail his coal in that city. He informed Browne that "Alabama coal is liked by many both for its cheapness & quality. We have none & would be glad to have some to supply the demand. We . . . would be able no doubt to sell a good deal of yours." Baker also reported depletions of his reserves due to recent shipments to Marion. By mid-February 1858, a warm spell caused coal sales to approach a standstill. Baker reported that the ACMC had depleted its stores, but that he had enough to continue operations due to reduced demand. A cold snap in early March revived the coal trade and actually expanded Baker's business. In spite of this upswing in the late winter and early spring, Baker and R. H. Kerr, his assistant, remained despondent. Predicting the end of the coal season, they reported coal sales of only one or two tons per day. Kerr complained: "The prospect is very gloomy as respects coal. . . . I have only sold . . . two hundred tons this winter. We could not get the coal when it was in demand, now when there is no demand we can get plenty." Still Baker and Kerr continued to perform their duties on Browne's behalf. They secured guarantees from the A&TR Railroad concerning freight rates and storage platforms, and Kerr made preliminary arrangements with an agent to supply 150 tons of coal to Howard College in Marion during the upcoming winter.[53]

As summer turned to fall, Browne witnessed an unprecedented expansion of his coal markets. Establishing relationships with new agents—Thomas K. Hooper in Uniontown and Eli Loveland in Marion—Browne set the stage for increased sales during the winter of 1858–1859. As he established his network of agents in these new areas, Browne seemed to gain focus on exactly how his business would develop. While in Selma, he dispatched the following detailed directives to Jasper T. Campbell who superintended the work at the Montevallo mines: "An order goes up from Union Town or Marion today for a load of [small screened] coal. . . . It must be good nut coal from that which lies on this end of the platform." Browne continued his precise instructions by stipulating that "some small lump coal may be put in if convenient—a little of it and but a little. . . . If any orders come from Marion & Woodville for lump coal, have them filled by the same train."[54]

Noting these developments, Baker notified Browne that he would no longer take a personal interest in Uniontown, Marion, Montgomery, or Mobile markets, but that his partner, a Mr. Perkins, would be willing to establish the river link between Selma and Mobile. Perkins had no capital of his own, but he would build and load the flat boats in Selma and handle the Mobile sales personally. Apparently, Perkins succeeded in establishing the Mobile trade. Within three months, Browne received a proposition from a Mobile firm offering to sell his coal. By year's end, Baker presented specific recommendations for further expansion of the coal market in Mobile.[55]

Market expansion constituted a two-edged sword for Browne. In early January 1859, Baker acknowledged receipt of two carloads of coal, but notified Browne that he was unable to deliver them to the appropriate customers. Apparently, demand far exceeded supply, and Baker found himself in dire straits in Selma. Concerning the coal cars, he informed Browne: "We rec[eive]d notice at once that unless we kept it [the coal] here, the people would seize it. . . . I cannot describe the excitement, if you were here you would go crazy. We have no more control over the coal than the 'man in the moon.'" Baker also lamented that "worst of all they refuse to pay . . . & some were mean enough to stop the carts in the streets, sent with coal to others, & appropriate it, & not let us know who got it, & the carts would not return, & so we lost about 4 tons."[56]

Furthermore, Baker sent an update on ten Uniontown accounts, informing Browne that all but one complained of poor quality coal and demanded reduced prices. Complaints included weight discrepancies, improper size and type of coal, handling and delivery problems, accounting incongruities, and payment-collection difficulties. Baker added: "When coal comes in all hands can hardly keep it straight. It is [hard] work to collect the dribs of bills." To further complicate matters, in mid-January, Baker wrote Browne from New York City. He intended to stay in New York for the remainder of the year to pursue contacts in the rudimentary coal oil industry. According to Baker, those prospects exceeded the business opportunities of Alabama by "50 to one." Consequently, Browne left Margaret at home to tend matters dealing with the mines and the railroad shipments while he moved temporarily to Selma to manage affairs there himself.[57]

Fortunately, Margaret had already become involved in the "family" business.

As early as October 1858, she had directed the loading and shipping of cars in Browne's absence. Ironically, Jasper Campbell suffered a cut heel only a few days after receiving Browne's dictates, so Margaret took charge while Campbell recuperated. About one month later, she informed her husband that Campbell's work schedule remained uncertain because his injury (the heel was cut to the bone) was more severe than originally thought. At Browne's direction, she relayed a daily report on the number of trams loaded and transported to the shipping yard. In addition to her dutiful reports, Margaret also related a plethora of problems seemingly inherent in this developing operation. According to her letters, lost days due to sickness, loitering and malingering, pulling trams by hand, and late-night loading combined to create continual challenges.[58]

Problems continued to mount, and by mid-January, Margaret exhibited little hope for better conditions. She informed her husband of Jasper Campbell's marriage and stated that he and his bride had moved five or six miles away. She stated: "I don't think he has any notions of going to work for you, though he told me on Saturday that he would go to work sure this morning." Ending the letter with her personal assessment, she concluded: "I don't think your mines are paying expenses now."[59]

Nevertheless, Browne's coal business continued to develop. His reputation spread considerably as evidenced by a letter from George Baker in which he concluded that the ACMC "are foolish in not putting their mines under your control. I know of no other man in Alabama, or out of it, that could work it successfully." Baker continued his flattery by dubbing Browne "the Father, the Gog & the Magog of the Coal trade."[60]

In late spring, Browne decided to return to Montevallo to manage production instead of dealing with retail matters. Shortly thereafter, he received word from the Mobile Gas Company that it desired to procure all materials from Alabama if practicable. Conditions continued to look favorable in late summer when Browne received word from Eli Loveland that he had closed the deal to supply 150 tons of coal to Howard College in Marion.[61]

A serendipitous opportunity also developed in the summer of 1859, initiated by J. F. Conoley of Selma. As events unfolded, Browne learned that Conoley represented P. J. Weaver and sought to negotiate the sale of Weaver's interest in their jointly owned coal lands. He invited Browne to participate in ongoing negotiations with "parties in New York." The "parties in New York" turned out to be none other than George O. Baker. Writing in mid-September, he informed Browne: "I have seen PJW [Weaver] at his hotel, & had hoped to hear from you before this with particulars. I want a plot of the mines as I wrote you. . . . I can no doubt be of great service to you." According to Baker, Weaver offered to lease his share of the coal properties. He continued: "If I should go back to Selma, he [Weaver] is coming down to see me in a few days & I hope I will hear from you in season to know how to talk. He will no doubt rely upon me to look after his interest, in case of a division." Apparently, Weaver was unaware of Baker's relationship with Browne. Consequently, Baker implored Browne to safeguard their communications: "Above all things, don't keep my letters, or leave them where they can blow out of doors. I would not have W think I am communicating with you on this subject or any account."[62]

Baker's duplicity continued as negotiations developed. Informing Browne of his dealings with Weaver, Baker instructed him concerning his "scripted" machinations. Characterizing Weaver as "a very suspicious man," he advised Browne to proceed with caution. Baker stated further that Weaver needed to remain unaware of any connection he had with Browne or the Montevallo mines. He assured Browne that he would return to Selma no later than the first of November and that, if all went according to plan, the transaction would be completed in due course. Again, he cautioned Browne: "But you & I must not be on too familiar terms where it can come to his notice. He is awful keen & will turn upon his heel on the least cause of suspicion."[63]

Baker did return to Selma as planned, and he assumed control of Weaver's holdings. He picked up immediately where he had left off earlier in the year, and his presence in Selma marked the culmination of a challenging, but profitable year for Browne. The 1859–1860 coal season looked promising, and many of Browne's business efforts were coming to fruition. Having struggled through nearly four years of sporadic operations, Browne and Baker had reason for hope as the year waned. Adjusting to seasonal markets, unpredictable production, fledgling transportation, and continual competition, these business partners persevered to establish the basis for a viable coal mining enterprise.

The fall of 1859 offered yet another development in the life of the Cahaba coal field. At the same time that Browne and Baker solidified their relationship, an English mining engineer appeared on the scene. Joseph Squire, working at different times for both Browne and for the ACMC, would revolutionize coal mining in central Alabama. Through practical experience, exploration, testing, charting, and mapping, Squire would emerge as the central figure in the development of the Cahaba coal field.

2
MINING AND MAPPING COAL
1859–1883

Joseph Squire may be the least prominent of the personalities involved in the development of the Cahaba coal field, but his role was central. Born the son of an English naval officer on November 24, 1829, at Rochdale, Lancashire, he attended school in England until his father's untimely death. Instead of pursuing a career as a naval officer as planned, he opted to work in a nearby coal pit. Squire stated that he "made every effort to learn to do any and every kind of work done in the pit." Drawn by the siren song of America's market revolution, he immigrated to the United States in 1849 and served a one-year apprenticeship as a machinist at the Peabody Furnace in Providence, Rhode Island. Ignoring cries of "gold!" that generated a rush to California, Squire traveled westward by rail to Pittsburgh and then by steamboat to the Kansas-Nebraska Territory. There he opened several coal mines to supply settlements and steamboats along the Missouri River. While based in St. Louis, Squire learned of high-quality coal and extraction problems in Alabama and moved to Montevallo in the fall of 1859. At this point, Alabama mining operations remained sporadic, with a handful of entrepreneurs exploring undeveloped seams. Upon his arrival, Squire found that disheartened miners had stopped work while awaiting the arrival from Wilkes-Barre, Pennsylvania, of a steam engine (for hoisting coal) powered by two thirty-inch cylinder boilers. According to Squire, the miners remained cautiously optimistic that the hoisting steam engine would lessen their labor and workload. He recorded further that "the scene was a picture of desolation such as I had never seen before." Relying on his knowledge of the coal mines of his native Lancashire, he sought to improve the techniques used by the Montevallo miners and thereby to increase production.[1]

Squire immediately encountered many of the trials experienced by mine owners in the antebellum South. In contrast to the situation he confronted in Alabama, mining operations in Pennsylvania and West Virginia mechanized long before those of the Deep South. For example, in the anthracite regions of Pennsylvania, five stationary steam engines were mounted in the Schuylkill fields by 1833; from 1839 to 1849, steam engine use for hoists and pumps increased fifteenfold. In the 1840s, demand for more and cleaner coal required extraction below the water level. This challenge increased the need for capital investments such as shafts, pumps, boilers, hoists, rolls, and breakers. Resisting widespread pessimism, Squire and various entrepreneurs sought to buck the trends that supported a traditional agrarian-based, anti-industrial economy.[2]

Most of the mines in the Montevallo area were owned by the Alabama Coal Mining Company (ACMC), and Squire contracted with company president John S. Storrs to resume extraction of the coal. In contrast to the prevailing rate of $10 per ton, Squire promised to mine lump coal (assuming continuous operation) and to load it onto rail cars for $2.50 per ton. Receiving a $900 advance from Storrs ($300 per month for the first three months), Squire formed a partnership with Alexander Anderson and John Whitehead to increase his capital. Increasing wages to $1 per ton, Squire recruited miners and opened a new entrance to the Irish Pit. His successful operation fostered resentment from a company superintendent named Donalson, however, and Squire determined to get away from Montevallo upon completion of his contract in 1860.[3]

Meanwhile, William Phineas Browne and George O. Baker continued to deal with numerous challenges. Perennial difficulties remained as they attempted to balance supply and demand. Quality control persisted as the primary criterion in competing with the ACMC. Floating freight rates and seemingly arbitrary regulations also kept transportation a major concern. Political turmoil and economic uncertainty added to feelings of futility. Such disruptions were exacerbated by slave unrest and by money shortages. Finally, both Baker and Browne experienced personal hardships that further complicated their attempts to establish a viable coal venture.

In January 1860, Baker informed Browne that the ACMC delivered coal in Selma at $7 per ton. This matched Browne's price ($6.50 on the yard, $7 delivered), but Baker projected sales of at least one hundred tons per week. Therefore, he remained concerned about supply from the mines and transportation of the coal by the Alabama & Tennessee Rivers (A&TR) Railroad. He reminded Browne that railroad officials held direct authority for allocating platform cars and thus should be treated kindly. He added: "They know we are only agents for you & think themselves second only to the savior of mankind & we must all bow accordingly."[4]

Baker also expressed concern regarding coal supply and competition from the ACMC. He pleaded with Browne to send more coal, referring to several paid orders that remained outstanding after three weeks. Too, he forwarded copies of advertisements circulated in Selma by the ACMC. One solicitation announced: "The Alabama Coal Mining Company is now prepared to deliver Coal in any quantity by cars or by casks, as ordered. Every car is weighed on car scales at Selma, and freights to that point paid by the Company." An accompanying newspaper article continued in the same vein: "The present is a good time to lay in your winter supply of coal. The Alabama Coal Mining Company, by the introduction of the most approved machinery, are now able . . . to supply demands to almost any extent." Baker sounded desperate when he informed Browne of ACMC attempts to expand its coal markets to Marion and Uniontown, and he assumed that the company's willingness to pay freight rates indicated successful sales. From Baker's perspective, any delay in delivering coal would bolster the ACMC's position in these new markets.[5]

Marion sales were bleak in the early weeks of 1860. Agent Eli Loveland, responding to a query from Browne, substantiated Baker's reports of insufficient coal supply. Rather than purchasing coal in Marion at $9 per ton, many of Loveland's customers ordered from Baker who charged an additional $3 per ton to cover transportation costs from Selma. Apparently, Browne commented on the absurdity of this arrangement, but Loveland countered by reiterating his dearth of coal. He reminded Browne that he had received only two carloads of retail in January and that he had to supply coal continually to Howard College to fulfill contractual agreements. He also relayed a message from a Mr. Gardener of the Alabama & Mississippi Rivers Railroad in which he offered to send cars directly to the Montevallo mines as long as they remained earmarked for Marion. According to Loveland, this arrangement would cut out the middlemen of Selma (both Baker and the A&TR Railroad), reduce the price of coal in Marion, and thereby increase sales in that area. On the other hand, if Browne insisted on managing prices and sales from a distance, then Loveland would terminate his agency no later than the first of April.[6]

Business did take a turn for the better in March 1860. Baker reported plenty of coal in Selma, and Loveland received seven carloads within a week. While in Mobile in late March, Browne received a letter from Margaret in which she reported eight trams of coal mined the previous day, and a half load of lump coal and one load of nut coal at the railroad platform. Success was short-lived, however. In mid-April, Browne received notice from Thomas A. Walker, president of the A&TR Railroad, instructing him to remove his Selma platform at his earliest convenience.[7]

Baker confirmed the news from Walker, having received verbal instructions from railroad official Rothrock. According to Rothrock, the railroad intended to build a freight house in the vicinity of Browne's platform and needed the space for a side track. Baker asked for special consideration, but the railroad company refused to make concessions, apparently basing its decision on an outstanding controversy with Browne. Baker did arrange to dismantle only a portion of the platform by the first of May, but Rothrock insisted that he be prepared to remove the remainder on twenty-four-hours' notice. Thus, albeit temporarily, Baker was forced to curtail coal shipments to Selma.[8]

Baker remained optimistic, however. He wrote Browne: "I hope your coal vein may get thicker instead of thinner as you get into it. [Five] years good luck would make you as rich as you could wish for." Alanson Saltmarsh of Cahaba confirmed this favorable outlook when he informed Browne that "I am very glad to hear that your prospects in your coal mines are encouraging." This air of optimism was well-founded. Loveland ordered four carloads in mid-May with prospects for more sales through the summer as customers began to make early arrangements for winter coal supplies. In addition, M. M. Bookes of Uniontown ordered forty tons of smithing coal. Even so, transportation remained the critical factor. With Browne in Richmond buying more slaves, Margaret reported an average of ten or twelve trams extracted per day, but only two carloads shipped. Daily production increased to fifteen or sixteen trams, but Margaret remained pessimistic about the labor force. She wrote her husband: "I hope you will purchase some Negroes before you return even if you should have to stay a week or two longer to accomplish it." She emphasized further the lack of efficiency among the slaves and her desire for a full-time resident overseer. Margaret asked for "one that understood the business and . . . that can manage Negroes for they need some one badly."[9]

As the summer wore on, some elements improved while others declined. For example, coal quality seemed to gain a significant boost. After seeing two carloads bound for Marion, Baker remarked to Browne that "if you send such coal as that, it will suit a prince." Demand remained high as well. Loveland reported from Marion that he had taken orders for nineteen carloads in addition to seventy-five tons required by Howard College. Orders were plentiful even though he expected no payments until November or December. Still, he encouraged Browne to increase production. Unfortunately for Browne, some of his slaves ran away, and, consequently, production decreased at this critical juncture. As historian Ronald L. Lewis reports, an increased reliance on slave labor accompanied the growth of an industrial base in the South. In the coal and iron enterprises of Maryland and Virginia, slaves formed the core of the labor force. In similar fashion, Browne's output dropped significantly when the number of slave miners declined.[10]

Moreover, Loveland informed Browne that the ACMC continued to send good coal and to fill orders promptly. To further complicate things for Browne, Baker wrote that "times are getting awful tight here & we cannot collect a dollar. . . . We never saw so much talk about hard times." Obviously, discussions of secession and states' rights had considerable impact on the overall economy in the fall of 1860.[11]

Nevertheless, demand for coal remained relatively high. Browne received orders from Uniontown, Cahaba, and Selma in addition to those already committed by Baker and Loveland. Baker's clerk also advised Browne that they had secured a new platform for shipments to Selma. Later, Baker noted that he still faced resistance from railroad and city officials. Despite their opposition, he continued to accept orders for coal and to conduct business as usual. Still, as cold weather approached and uncertainty persisted, he requested specific instructions from Browne. Labor problems continued to plague Browne, however. The pursuit of runaway slaves consumed his energies as he communicated with individuals as distant as Rome, Georgia. In addition, he brought suit against slave dealers in Richmond, Virginia, due to health problems experienced by a recently purchased slave. These problems persisted through the end of 1860 as countersuits and financial liabilities complicated the litigation process and continued to distract Browne from his mining pursuits. Thus, Browne encountered many of the frustrations experienced by colliers in Maryland and western Virginia during the early nineteenth century.[12]

Politics also demanded much of Browne's time. Having remained active in state and local politics since his tenure in the state legislature, Browne found himself at the pulse of central Alabama activism as the presidential election of 1860 approached. In late October, he invited Alabama governor A. B. Moore to address a political rally scheduled for the Montevallo area. Responding favorably to Browne's invitation, Moore stated that he believed John C. Breckinridge to be the only candidate capable of defeating Abraham Lincoln. A Republican victory, he claimed, would destroy the constitutional rights of every citizen in the slaveholding states. Moore concluded: "I deem it my duty, as well as the duty of every man to do all that may be done by talking and voting to defeat this election and thereby save the Constitution and the Union."[13]

As political issues developed, economic woes continued. Baker complained of being unable to collect as much as $100 to pay a freight bill. Even though he had more than $6,000 in assets, he could not collect any cash for coal, crops, or services. This money shortage prompted him to warn Browne to "look out for hard times. They are full upon us, & seem [to] be getting worse daily. . . . Everybody has quit business to talk secession. . . . We hope to see a change soon, either to break up the Union or to stop talking about it, & then people will pay up."[14]

Orders continued in spite of political and economic uncertainty. As cold weather became imminent, customers sought to ensure that they had sufficient heating coal for the winter. Activity increased for Browne as well as for the ACMC. Baker commented that "the competition will now be in the quality of coal. We are all ready to receive coal at the Selma yard. . . . Send us good lump coal to start up on." Furthermore, he encouraged Browne to sustain his supply of coal lest ACMC agent Ellsberry absorb part of their Selma market. Realizing that regaining customers once lost would be difficult at best, Baker coaxed Browne by deeming it "unfortunate that we are out of coal. Ellsberry gets all the trade, but we sell all we can get."[15]

Baker's importuning had little immediate effect. Browne's slack coal could not compete with the comparatively large lumps extracted by the ACMC. Ellsberry retained the upper hand in Selma sales due to better quality, and he shaved $1 off the price of fine coal, thereby undercutting Baker's market. Baker's outstanding accounts had increased to $8,000, but collection of debts remained an acute problem. Even so, he remained optimistic. His last report of the year ended with an upbeat message: "We now have a new cart, new mule, new Negro & everything 'tip top' for the coal trade. All we want is plenty of good, 'black diamonds,' to make the investment pay."[16]

As 1861 dawned, secession loomed dark over Alabama. Declaring itself a "free, sovereign, and independent state," Alabama seceded from the Union and joined the Confederacy on January 11. Three months later, Confederate forces fired on Fort Sumter in Charleston Harbor, and the American Civil War commenced. Secession and war generated uncertainty and turmoil, however, and political and economic pressures weighed heavily on both Browne and Baker in the early months of 1861. Working feverishly to continue production, maintain sales, and collect outstanding debts, the pair struggled to preserve their enterprise. After receiving a letter pervaded with criticism from Browne, Baker responded: "We have sen[t] out good coal, but you have sent a good deal of bad coal by carloads. We have worked off any quantity of fine coal for you, that nobody else would have got rid of at one dollar less per ton." After vehemently countering Browne's accusations, he commented: "We think it woul[d] be to your interest [t]o make new arrangements.... We like to be stirred up a little when we really deserve it, but you come down upon us a little too harsh."[17]

Browne did not make other arrangements, and Baker soon became caught up in the wartime excitement. Apparently residents of Selma were in an uproar. After a week of war news, he wrote Browne: "Everything [is] upside down generally." In spite of the turmoil generated by war, Baker, Browne, and many other entrepreneurs looked beyond the disarray and identified an opportunity for significant economic success.[18]

By early 1861, Browne owned 400 acres of coal lands, a half interest (with P. J. Weaver) in an additional 1,200 acres, a 200-acre home place near Wilton, a bloomery or forge named "Brighthope" in Bibb County, and eleven slaves valued at $10,000. But with the vote for secession and the outbreak of the Civil War, he also faced a declining coal market, labor shortages, and interrupted river transportation. Consequently, he hired Joseph Squire as superintendent of the Browne Pit for twelve months. Squire, having completed his contract with the ACMC in the fall of 1860, had headed for Australia. En route, he had received word of trouble with his mining interests in the Kansas-Nebraska Territory. Returning to the United States, he headed west to assess the situation himself. After determining those operations to be unsalvageable, he abandoned the Missouri River mines and returned to Montevallo. In February 1861, he contracted with Browne to open a hundred-foot slope into the Montevallo Seam using steam-powered hoisting technology. During the ensuing year, Squire boarded at Browne's house in Wilton, and Browne furnished "a nag" for Squire to ride to the mine and back.[19]

Shortly after Squire began working for Browne, the American Civil War erupted at Fort Sumter. Apparently the news traveled quickly; Squire's journal entry for April 19 states that he "went to M[ontevall]o to hear the speaking on the war." Also, "Neg[roe]s John & Fleming ran away today." In fact, the war would have a profound effect on Alabama's coal-mining industry. Ethel Armes reports that "the large majority of railroad, coal, and iron men had voted against secession. Only when the issue became irrevocable did they stand by the State." Once formed, the Confederacy suffered from a dearth of ironworks. Richmond, Virginia, claimed the only cannon foundry and a rolling mill; other rolling mill sites included Shelby, Alabama, and Etowah (near Cartersville), Atlanta, and Columbus, Georgia. War demands stimulated growth in coal and iron industries throughout the southern states. Thomas S. Alvis came to Alabama from Virginia to establish the Confederate arsenal at Selma and another rolling mill at Helena. West Virginian Richard Fell and associates R. W. Cobb, B. B. Lewis, Richard Fell Jr., and Charles Albert Fell contracted with the Confederate ordnance department to construct a rolling mill at Brierfield.[20]

Due to access provided by the newly constructed South & North Alabama Railroad, the Cahaba field became the primary source of coal for the Confederate government. Armes states that most of the coal supplied to the Confederacy came from six counties—Tuscaloosa, Jefferson, Walker, St. Clair, Bibb, and Shelby. "Every train and barge load was concentrated at Selma and distributed from that station to Montgomery, Mobile, and other points." With Selma serving as a "clearing house," Browne's and Baker's coal enterprise promised to reap the benefits of a wartime economy.[21]

On the other hand, economic success would not come easily. In spite of continual orders and deliveries of coal, cash remained scarce. Therefore, even though Browne and Squire could extract quality coal and Baker and Loveland could deliver it, few individuals were able to pay cash upon demand. Just two weeks after war ensued, Baker explored the possibilities of supplying coal to Mobile in the event that the Union successfully blockaded Mobile Bay. However, a retail price of $10.50 to $11.00 seemed extravagant in light of the money shortage. Furthermore, news from Selma described the overall effect of the war on Alabama's economy. Baker informed Browne: "Failures are occurring here frequently. . . . The legislature will no doubt pass a general bankrupt law after the war & leave us all minus

& broke. So you had better not be too liberal with your coal sales." Two days later, he recorded losses due to bankruptcies of $900 and $2,000, respectively. In light of widespread business failures, Baker concluded: "It is not safe to trust one's Grandmother, but we are getting everything on a good war footing so as to give the North all they want."[22]

Squire's was not a panacean existence either. His journal entries reveal a demanding work schedule of day and night shifts that stretched from predawn to dark six days per week. He supervised five slaves in bailing water from the pits, opening ventilation shafts to counter stale air, building a bridge over a mining ditch, and cutting crossties and laying track for coal trams. He also mentioned that he "was provokingly bothered with Browne about the firming the track [*sic*] and equalizing the ties." In addition to dealing with Browne's domineering personality, Squire filled his days by repairing pick handles, sharpening saws, and bracing the mine roof with timbers. Also, in an effort to increase efficiency by reducing time and effort, he "fixed the dirt dump so as to enable the cars to be emptied without shoveling." Squire worked incessantly to establish an effective mining system, but his miners filled as many trams with dirt as with coal in these initial weeks. He also rebuilt the blacksmith shop after a fire destroyed that structure, and he assisted in the construction of a new house for Browne's family.[23]

These myriad tasks and the long hours prompted Squire, in desperation, to consider escaping to the West Indies, Mexico, or South America. He remained true to his agreement, however, and successfully negotiated a compensation settlement with Browne. Browne would pay $150 for Squire's work to June 1 (approximately three months) and then $2 per day thereafter. In the ensuing months, Squire seemed to be especially sensitive to interruptions in production. For example, he informed Browne that working double shifts proved ineffective. Squire stated that "the servants rely on stopping work at the stated time whether the complement of work is done or not." Also, coal production ceased for two days in May while his workers hauled ninety-six cars of water from the pit. Later, on August 13, he wrote: "This day was full of accidents. First, before sun-up, John let the cars run over him. Second, William got stung near the eye with a hornet. Third, they let the water in the mine while I was changing clothes and let [the] mule run away." In the same week, Squire commented on disciplining various slaves by whipping or by giving "him what he deserved for it." About one month later, he "was aggravated . . . very much with the carelessness, lying, and meanness of the Negroes. One let the mule get loose and others would not work when I was out of their sight."[24]

Meanwhile, Baker complained of floating freight rates. Railroad official Millington established a fixed rate for cargo weighing up to eight tons, but rates increased significantly for cars with heavier loads. Moreover, Browne and Baker continued to complain of too much dirt in the coal, and Baker even visited the mines in early August in an attempt to improve quality. Consequently, Squire's purview extended from mining the coal and filling the trams to conveying the trams to the platform and loading the rail cars for transport. In spite of Squire's frustrations, his hard work paid off. From mid-June through mid-September, Squire and his workers extracted more than 1,500 tram-loads of coal from Browne's mines.[25]

Coal production alone could not boost the economy, however. By late September, Baker concluded that "coal is not the best thing to invest in. . . . I would sell anything now for money to operate with. Everything is ten times worse than it was 10 or 15 days ago." Bringing the destitution to a personal level, he added the following postscript: "Our diet at home is sweet potatoes & milk. As for ham & coffee they are 'among the things that were.' We have settled down to starvation principles." Conditions continued to worsen as Baker reported dwindling supplies of groceries and all but three stores closed on Water Street. He added that "there is no coal selling, & nothing doing of any kind." Apparently, this pessimistic tone incited Browne's ire inasmuch as Baker's next communication confirmed that "'all business connection between us ceases,' according to your determination." The rift proved temporary, but a few days later, when Baker reported that two or three children died daily in Selma, he recommended again that they "give up the coal." He determined that "we cannot give it the attention it needs, without help, & will not pay enough to pay help & risk of getting an honest fellow to do it."[26]

Browne may have heeded more of Baker's advice than he admitted. Early in 1862, he received a letter from William Wirt of Atlanta declining Browne's proposal to form a partnership for mining coal. This flirtation with the idea of reducing his control over the mining venture was short-lived. Just four days after the arrival of Wirt's negative response, Browne heard from John W. Lapsley of Selma concerning the Confederate Army's intent to establish a foundry and arsenal in that city. This enterprise would create a sizeable and continual demand for coal and iron, and Lapsley instructed Browne to "write me immediately and state the price at which you will furnish here . . . all the coal the government may require for the foundry for a given period [two or more years]."[27]

At last, Browne and Baker realized the breakthrough that they longed for. Unfortunately, the Confederate government experienced many of the same cash-flow problems as Browne's individual customers did. Browne spent more than a year petitioning the Confederate Naval Department for payment. Finally, in May 1863, he received a response from Julius A. Pratt of the Confederate Naval Station in Selma. Offering no resolution, Pratt merely appealed to Browne's patriotism and sense of duty: "These are trying times, and everyone should lend a helping hand. If you have been wronged, you have a governor and friends to appeal to. At this time the government requires the coal more than ever."[28]

Baker offered to assist, but he expressed little hope for a quick settlement. Encouraging Browne to send him a power of attorney, Baker commented that the document "will have to be perfect in all its parts to pass muster in the circumlocutionary red tape departments of government. I would rather give them a small bill than try to collect it." Apparently Browne responded to the bureaucratic delays in his typical caustic manner prompting Baker to send him a friendly warning. Baker urged Browne to "deal more gently with the Naval powers 'that be' if you want smooth sailing. . . . [I]f you have made a bad contract you must throw yourself upon their clemency or they are sure to make plenty of trouble for you."[29]

Browne had other concerns as he wrestled with the Confederate government. While working for Browne, Squire negotiated a new contract with Storrs to resume superintendence of the ACMC mines in March 1862. This arrangement afforded Squire $40 more per month for his efforts. He quickly gained the approval and confidence of the board of directors, but his workload increased considerably. By the end of 1862, he managed three miles of railroad; supervised the Irish Pit; mapped developments of the Irish Pit, Shaft Seam slope, and the Montevallo Seam; and tended the company store from noon until night on Wednesdays and Saturdays.[30]

Browne experienced sickness and sorrow in the fall of 1862. While returning from a business trip to the Confederate capital in Richmond, he contracted a "soldier's fever" that incapacitated him for nearly two months. Shortly after his return home, Margaret fell ill with the same fever; she died five days later. In addition to inflicting physical and emotional suffering, Margaret's death in October 1862 compounded Browne's business problems. Although he continued to pursue his mining operations, he faced many of the same issues that had plagued him for years. His new superintendent, P. M. Fancher, wrote of slave schedules, track and crosstie maintenance, and tram repairs. Browne's focus shifted increasingly to his fledgling ironworks at Brighthope, and he directed most of his energies to enhancing that facility. As a result of all this battering, on July 16, 1863, Browne sold his coal mining interests to the Mobile and Selma Coal Mining Company (MSCMC). This newly formed enterprise consisted of a partnership between George O. Baker and Isaac D. Spear of Mobile. The partners paid $137,000 for the mines and lands and then contracted with Browne to supervise construction of a branch road to facilitate mining in the Irish Pit.[31]

The transaction almost fell through a fortnight later when Baker informed Browne that "Capt[ain] Pratt of the Naval department advises me that 'every d_____ ton of coal from your mines will be seized.'" He related further that he had sold one hundred tons to the A. Sellers Naval Foundry, but that Browne should consign the rail cars to the MSCMC or to Baker himself. If they could avoid confiscation by Pratt, then they could recoup Browne's initial expenses totaling $12,000. Browne supported Baker's plan, and ownership of the Montevallo mines transferred to the MSCMC.[32]

Baker, Browne, and the ACMC were not the only parties interested in exploiting the resources of the Cahaba coal field during the Civil War. In fact, many entrepreneurs took advantage of the Confederacy's policy exempting government workers from military service in the field. By establishing a contract with the Ordnance Department, anyone with at least twenty slaves could supply coal to the Confederate government and thereby avoid military service. Still, regulations stipulated that all affiliates drill their employees daily to train them for wartime emergencies. Thus, in addition to the mines owned by Browne and the ACMC, other operations dotted the Cahaba region. William Gould, characterized by Armes as a miner, prospector, geologist, operator, and discoverer, managed the Helena mines of Monk, Edwards, and Company. Born near Glasgow, Scotland, in 1830, "Uncle Billy" came to the United States in 1852. Arriving in Alabama two years later, Gould gained recognition as "the pioneer mine hunter and tracer in the state" and as the first person to make coke from Alabama coal. Partnering with Charles and Fred Woodson, Gould established a coal mine and several coke ovens at the confluence of Buck Creek and the Cahaba River (near present-day Helena). During the Civil War, the partners shipped seventy-five tons of coal and coke per day to Selma, but Wilson's Raiders demolished the workings in March 1865, setting three thousand tons of stockpiled coal afire.[33]

In addition to the Gould and Woodson mines that were located near the Cahaba River, the Dailey Creek Basin contained numerous temporary drifts, or "bomb proofs." These smaller mines represented the efforts of citizens from Mississippi, refugees forced by Union forces to flee their homes. One account, related by Frank Fitch, tells the story of the Thompson family of Six Mile in Bibb County. Leaving their home in Hinds County, Mississippi, after Union forces burned their house and their cotton crop, brothers N. H. and Lewis Thompson moved their families and slaves into central Alabama. Contracting to mine coal for the Confederate government, Lewis and his son, Julius, opened the Lower Thompson mine, and N. H. opened the Upper Thompson mine. Both mines tapped what became known as the Thompson Seam, and the Thompsons transported their coal over a twelve-mile route to Ashby, a station on the Selma, Rome Road, and Dalton Railroad. Fitch provided his own description of the operations: "No explosives were used . . . but the pick and bar did the work, and the cabs were hauled to the surface with mules. . . . It was a severe life for them for they had left their comfortable Mississippi plantations and were . . . crowded into makeshift huts and shanties."[34]

Other businessmen took advantage of the demand for iron and the interrelation of coal mining with iron production. The Red Mountain Iron and Coal Company, organized by Frank M. Gilmer Jr. and John T. Milner in 1862, represents the quintessential example of this form of enterprise. As officials of the South & North Alabama Railroad, Gilmer and Milner merely expanded their transportation interests to incorporate coal and iron. In fact, Gilmer, as president of the

South & North, had commissioned Milner, the company's chief engineer, to survey the area from Montgomery to Decatur in December 1858. Milner's report, submitted approximately one year later, advocated a route through the Cahaba coal field to tap the mineral resources of central Alabama. Once established, the railroad would provide the necessary link between the iron ore of Red Mountain and the coal of the Cahaba field to facilitate the manufacture of iron. Upon completion, the South & North route would connect the Alabama River Valley at Montgomery with the Tennessee River Valley at Decatur. Also, this critical tie would provide access to markets in other parts of the state and beyond. Therefore, Gilmer and Milner chose to establish Oxmoor furnace near the railroad and at the foot of Shades Mountain.[35]

The stockholders of the South & North Alabama Railroad commissioned the Red Mountain Company and hired Moses Stroup, builder of the Tannehill furnaces, to oversee operations at Oxmoor. With Frank Gilmer's brother, William, as president, the furnaces went into blast in the winter of 1863 coincidentally with the opening of coal mines at Helena. At this point, the South & North Alabama Railroad merely connected Oxmoor with Calera on the A&TR Railroad and was described as a "patchwork line, 'every sort and kind of rail from 60 pounds T to 30 pounds T and strap rail and stringer!'" Even so, this haphazard transportation link provided access to Selma and enabled Gilmer to negotiate a contract to supply iron to the Confederate government. Although producing fewer than ten tons per day, the Oxmoor furnace furnished the iron that was crucial to the manufacture of Parrott rifles, shot, and shell.[36]

By the spring of 1864, Milner believed that the Red Mountain Company could establish itself firmly within the coal and iron industry of Alabama along with the potential expansion of operations throughout the Deep South. Estimating that the Helena veins contained six to eight million tons of coal, he predicted that, with a dedicated labor force (i.e., slaves) and completion of the rail link to Montgomery, the company could extract one thousand tons per day for more than twenty years. Milner admitted to Gilmer that it seemed futile to make long-term projections while in a state of war. On the other hand, he looked optimistically at the broader world of commerce beyond the war. Projecting the development of coal markets along the Gulf Coast, he predicted that "Alabama coal must . . . come from the Cahaba River, as the nearest coal fields to the gulf. . . . You therefore have a monopoly, or a perpetual bonus, to that extent."[37]

Things did not develop as Milner anticipated. Coal mining and iron production ended abruptly in the spring of 1865 due to the devastating cavalry raid by Union forces under General James H. Wilson. Armes reports that "the burning went on and on—and beyond that burning city [Selma] smoked the ruins of Oxmoor, Irondale, Tannehill, Brighthope, Brierfield, Shelby, and all the rest—the coal and iron business of Alabama, quieted now, it seemed forever." In summarizing the devastation left by the war, she observed that "every plant in Alabama had been silenced by Wilson's hand, and the State's coal and iron business . . . had been burned to the roots. Fully two thirds of the shareholders in the mining and furnace companies who survived were ruined in their personal circumstances."[38]

This assessment rang true for William Phineas Browne. He and Baker, trying to resume mining operations near Montevallo, experienced new problems in dealing with free labor. In June 1865, Baker forwarded a Freedmen's Contract of Labor along with thirteen workers, informing Browne that "the contract must be registered with the Provost Marshall & it is then binding." Apparently, Browne adjusted to the postwar arrangements; in 1866 he negotiated a two-year lease with Baker to reestablish his personal control over the mines. However, the following spring, Baker informed Browne that sale of the mines was imminent. Browne responded by submitting a draft drawn on the company's account. This action raised Baker's ire and prompted him to send the following missive to Browne: "You have received $3228.16 on this joint coal business, & we have not had a dollar, & you will readily perceive that there is some difference between Three thousand, two hundred & twenty eight dollars & nothing at all."[39]

Ultimately, a court order forced Browne to forfeit all claims to the coal mines and lands then in possession of the MSCMC. Seemingly in desperation, Browne offered his services to R. M. Moore, superintendent of the Cahaba Coal Company and recent lessee of Baker's mines. Moore responded curtly: "I do not see how we could well operate the coal mine together as my policy is to give any business personal attention while your plan is to operate on your men from your residence." Seemingly adding insult to injury, Moore stated matter-of-factly: "You have had a larger experience than myself, but . . . I think your health has been too much impaired to give the coal the attention it would demand." Thus Browne, the true pioneer of the Cahaba coal field, faced failure and rejection in the year preceding his death on January 13, 1869.[40]

In contrast, the war and its aftermath were kinder to Joseph Squire. John S. Storrs, president of the ACMC, died in early 1863 thus spawning a period of transition. In January 1864, a new group of investors purchased the company and changed the name to the Montevallo Coal Mining Company. Hiring Squire as mining engineer, company officials "left the fixing of my yearly salary to myself." Squire demanded and received the same salary as the superintendent and continued to supervise mining operations throughout the summer. In August 1864, in the wake of Union cavalry raids throughout Alabama, Squire assisted a group of Montevallo citizens in protecting women and children along the Coosa River from a gang of robbers. These thieves apparently preyed on families whose husbands and fathers were "at the front," and Squire suffered two bullet wounds in his left leg during this humanitarian effort. Invalided for several months, Squire received numerous acts of benevolence from company officials. President Thomas J. Portis provided a horse; General Superintendent George Figh supplied a cook; and Secretary and Bookkeeper Benjamin Davis "was a good Samaritan to me in so many ways that if I had been his brother ever [*sic*] so he couldn't have done better." These men obviously understood that Squire had become the indispensable authority on coal mining in the Cahaba field.[41]

Acknowledging Squire's value to mining operations, the Montevallo Coal Mining Company offered him control of $12,000 worth of stockpiled coal in May 1865. According to the proposal, Squire was to sell the coal for working capital and then repay the company in installments. At the same time, Browne offered Squire a half interest in his Baker Mine in exchange for supervisory and engineering services. Squire declined both offers. In fact, the Civil War's end created a drastic decline in the demand for coal, so Squire took an eighteen-month break for further study of mine engineering, geology, mineralogy, and coal and iron ore properties and formations. His hiatus marked the end of Browne's pioneering efforts and established another milestone in the development of the Cahaba field. War's end and Browne's death opened the way for Joseph Squire to establish himself as the preeminent authority on central Alabama coal.

In October 1866, Squire contracted with his former Montgomery-based employers to conduct a survey of the Montevallo Basin, the Lolley Basin, and the Overturned Measures, all in the vicinity of the Montevallo mines. While surveying this southeastern portion of the Cahaba field, Squire boarded at Isaac Taylor Tichenor's home in Montgomery and mapped his findings. Tichenor was pastor of Montgomery's First Baptist Church. He had served temporarily as a chaplain for the Seventeenth Alabama Regiment at the Battle of Shiloh (April 6–7, 1862), then returned to First Baptist in January 1863. Shortly thereafter, he helped organize the Montevallo Coal Mining Company and became its president in 1867. That same year, Tichenor financed board, traveling, and living expenses for Squire while he surveyed the southwestern corner of the Cahaba field. Tichenor biographer Kimball Johnson states that "Tichenor was among the first in Alabama to mine coal scientifically with the use of steam machinery. During these years Tichenor became closely associated with Joseph Squire, an Englishman, with whom he made prospecting trips through the State of Alabama and assayed the State's mineral and agricultural resources."[42]

After resigning his pastorate in January 1868, Tichenor devoted his energies to original research related to the development of Alabama's natural resources. Convinced that wealth and population combined to build the power of a state, he sought to reorganize various industrial interests to promote prosperity. Still boarding at Tichenor's house when not performing fieldwork, Squire provided a

Fig. 3. Isaac Taylor Tichenor, Baptist minister and advocate for the development of Alabama's natural resources. (Courtesy of Auburn University Libraries Special Collections and Archives.)

specific account of their relationship. His autobiography indicates that he loaded Tichenor's furniture and household goods onto a steamboat at Montgomery on October 27, 1868. Docking downriver at Selma, Squire transloaded the goods from boat to train and proceeded by rail to Calera and then on, via the South & North Alabama Railroad, to Siluria. From there, he transported Tichenor's belongings by wagon to the Blazingame Nabors house at Big Blue Spring (somewhere between Siluria and Helena in Shelby County). Squire "set up housekeeping" for the Tichenors, finishing just before the family arrived from Montgomery.[43]

Squire and Tichenor worked together to explore and define the Cahaba coal field, but the deaths of a two-year-old son in August 1869 and of his third wife a month later prompted Tichenor to return to the ministry. After serving for about a year as pastor of First Baptist Church in Memphis, Tennessee, he accepted the presidency of the Agricultural and Mechanical College of Alabama (present-day Auburn University) in the spring of 1872. Receiving his appointment from the board of directors with the approval of the state legislature, he served in that capacity for ten years. Johnson notes that "during President Tichenor's last years at Auburn the college assisted in the development of the iron industry of the Birmingham district." Selected by the Southern Baptist Convention as corresponding secretary of the Atlanta-based Home Mission Board, Tichenor assumed those duties in July 1882.[44]

Meanwhile, Squire continued to pursue mining efforts throughout the Ca-

Fig. 4. South & North Alabama Railroad Bridge No. 72. Joseph Squire's house is in the foreground; the town of Helena is in the distance. (Courtesy of Ken Penhale, Helena, and the University of Montevallo Archives.)

haba field. For example, in the spring of 1869, he journeyed to Pittsburgh and other parts of Pennsylvania to promote investment in Alabama minerals and railroads by northern capitalists. Even though he offered land at $3 per acre, this trip proved generally unfruitful. He returned to Montevallo where he resumed duties as superintendent and mining engineer of the mines. Resigning in February 1870, Squire moved to the Buck Creek area near Helena where he built a house and office. He worked as superintendent and mining engineer for Newman and Company Mining Operations along Buck Creek until the company forfeited its lease to the Red Mountain Iron and Coal Company. After an unsuccessful attempt to tap the Conglomerate Seam himself, Squire agreed to become superintendent and engineer for the Red Mountain Company at a salary of $100 per month. Beginning this work in October 1871, he built a tram bridge over Buck Creek and opened drifts into the Little Pittsburgh Seam running along the South & North Alabama Railroad embankment.[15]

Fig. 5. Truman H. Aldrich, coal-mining entrepreneur and cofounder of Pratt Coal and Coke Company. (Ethel Armes, *The Story of Coal and Iron in Alabama* [Birmingham: The Chamber of Commerce, 1910], 268.)

In May 1872, Daniel Pratt and Henry F. DeBardeleben, who had just bought a controlling interest in the Red Mountain Company, contacted Squire about continuing supervision of the Helena mines and also asked him to survey the Oxmoor Furnace to the north. He located additional sources of red iron ore, built a tram road from the ore mines to the furnaces, and notified residents in the houses there that they must vacate within a month because the company would need those structures to house its laborers. In that same year, Squire recorded that "Mr. T. H. Aldrich bought the Dave Smith interest in the Baker Mine at a point . . . about two miles a little north of west of Montevallo." Truman Aldrich began operations in 1873 and developed a mining community that would later bear his name. These affiliations with DeBardeleben and Aldrich, the two major developers of the Cahaba field, placed Squire at the pulse of coal-mining operations in central Alabama. In addition, relationships with Tichenor and State Geologist Eugene A. Smith further identified Squire as the preeminent authority on Cahaba coal.[46]

After graduating from Rensselaer Polytechnic Institute in 1869, Truman H. Aldrich left his native New York and moved to Alabama. Although trained as an engineer, he entered the banking industry in Selma before gaining interest in the coal mines around Montevallo. His brother, William F. Aldrich, came to Ala-

Fig. 6. William F. Aldrich, coal-mining entrepreneur and developer of coal town at Aldrich. (Courtesy of Aldrich Coal Mine Museum.)

bama from New York in 1874 and joined him in forming the Montevallo Coal and Transportation Company. The brothers worked as partners until Truman sold his interest to William in 1880. Seeking to increase his holdings and to facilitate further development of the Birmingham District, Truman Aldrich joined with James W. Sloss and DeBardeleben to form the Pratt Coal and Coke Company in 1878. Squire asserted that "this first effort of Mr. T. H. Aldrich in 1873, like the first effort of Mr. Henry F. DeBardeleben in 1872, did not benefit future generations near so much as the combined efforts of the two in 1878." Thus, Aldrich employed Squire at various times to survey Cahaba and Warrior lands for his purchase and development.[47]

From May through July 1877, Aldrich assisted Squire in conducting a geological survey of the southeast corner of the Montevallo Basin. This area included portions of the Baker drift, one of the mines that had been opened in 1861. In the last months of 1877, Squire worked for Sloss's Eureka Company, conducting an underground survey of the Wadsworth Seam near Tacoa Depot. Returning to Aldrich's employment the following spring, Squire searched for the best location for a large coal mine within six miles of the South & North Alabama and the Alabama Great Southern Railroads. DeBardeleben combined efforts with Aldrich in May, and Sloss joined their collective venture in July 1878.[48]

Fig. 7. James W. Sloss, coal-mining entrepreneur and cofounder of Pratt Coal and Coke Company. (From Ethel Armes, *The Story of Coal and Iron in Alabama* [Birmingham: The Chamber of Commerce, 1910], 128.)

Fig. 8. Henry F. DeBardeleben, coal-mining entrepreneur and cofounder of Pratt Coal and Coke Company. (From Justin Fuller, "Henry F. DeBardeleben, Industrialist of the New South," *The Alabama Review* [January 1986], 7.)

Hailing from north Alabama, James W. Sloss had earned his reputation as a merchant and plantation owner in Limestone County. Developing an interest in railroads in the 1850s, he helped negotiate the merger of the Louisville & Nashville and the South & North Alabama Railroads. This cooperative effort provided access to the raw materials of Jones Valley and contributed immeasurably to the development of Birmingham. Taking advantage of the natural resources within close proximity, Sloss determined to produce iron and founded the Sloss Furnace Company.[49]

The dominant figure in the development of the Cahaba coal field was Henry Fairchild DeBardeleben. Describing himself as a "piney rooter," a lean type of hog that roots into the earth searching for sustenance, he seemed to prefer the bold financial gamble to routine labor. Characterized by historian Justin Fuller as the epitome of the New South industrialist, DeBardeleben developed an audacious approach to investment. After his father's death in 1850, he became the ward of Daniel Pratt and eventually married Pratt's only child, Ellen. Inheriting most of Pratt's wealth in 1874, DeBardeleben sold his interest in Pratt's Eureka Company to Sloss for $160,000.[50]

These three Alabama capitalists gave definition to the "New South Creed" later espoused by Henry Woodfin Grady, Daniel Augustus Tompkins, Walter Hines Page, and others. Grady, editor of the *Atlanta Constitution,* outlined the key components of his vision to the New England Society of New York in December 1886. In contrast to the Old South of antebellum days, Grady's New South included industrialization, diversified and efficient agriculture, widespread education and vocational training, sectional peace, and racial harmony. In fact, textile, tobacco, lumber, phosphate, and sugar industries experienced significant economic growth in the late nineteenth century, and southern coal production increased tenfold from 4.6 million tons in 1875 to 49.3 million tons by 1900.[51]

After Aldrich, DeBardeleben, and Sloss formed the Pratt Coal and Coke Company, Squire began testing lands northwest of Birmingham to establish the Pratt Mines. He opened Slope No. 1 in October 1878, and a team of surveyors staked out the Birmingham & Pratt Mines Railroad a month later. With this transportation link to the Birmingham District, Squire continued his surveys and opened Shaft No. 1 (about one mile from Slope No. 1) in December. Throughout the following year, Squire worked primarily for DeBardeleben and Sloss, surveying coal in the Cahaba, Warrior, and Coosa fields for the Pratt Coal and Coke Company and the Eureka (or Oxmoor) Company, which they owned separately.[52]

In addition, Squire conducted a survey of the Big Warrior River from Old Warrior Town, site of an old Indian village, to Tuscaloosa from August through mid-December 1879. Working with State Geologist Eugene Smith and Professor Henry McCalley under the direction of Major Andrew Damrell of the Army

Corps of Engineers, Squire established excellent rapport with his fellow surveyors. Moreover, Smith recognized Squire's primary role in the early development of the region. One of his first letters reveals that he recommended Squire to Charles E. Hall of Philadelphia as a man "thoroughly competent to study up the magnetic iron ore beds of Alabama and Georgia."[53]

After eighteen months conducting iron ore surveys of McAshan Mountain and West Red Mountain for various investors, Squire returned to coal surveys. Following streambeds and examining outcroppings, he recorded the depth, width, and composition of various strata. Comparing exposed sides of eroded areas, Squire developed cross-sectional sketches that identified veins of coal as well as over- and underlying layers. He examined approximately 27,000 acres of coal lands near Dudley, a station on the Alabama Great Southern Railroad in Tuscaloosa County. This survey—commissioned by A. C. Danner and Aldrich, top officials of the Southern Mining and Transportation Company—continued from June 20, 1881, to October 9, 1882. At that time, Squire resigned from the company "for the purpose of making an extended trip to New York City."[54]

Squire returned to Helena ten days later with his new bride, his marriage surprising all of his Alabama associates. Aldrich and W. C. Thompson, secretary and treasurer for the Southern Mining and Transportation Company, conveyed their congratulations early in December. Thompson wrote: "hoping that the presence of a wife will prevent your having further attacks of rheumatism, or of being spilled out of wagons, and wishing you years of health, happiness, and prosperity." Apparently, Squire had an accident that prevented him from returning to work immediately, but he was able to take six months to enlarge and improve his Helena home.[55]

In January 1883, Aldrich entreated Squire to evaluate operations at Dudley. He stated that things were in poor shape and that "there is an absolute necessity for something to be done and done quickly." Squire promised to do so by March 12 and requested a loan from Aldrich for further home improvements. Aldrich provided the necessary funds and further bemoaned the poor state of operations at Dudley. He informed Squire: "I am very much exercised about our coal seams. It looks as if DeB[ardeleben] had out generalled us and we had purchased the poorest part of the field. Am decidedly blue about it." Squire informed Aldrich on March 4 that rain and frost would delay his trip to Dudley by ten or twelve days. He had experienced difficulty in getting building materials for his house, was plastering walls himself, and his wife had been sick for more than a week. The following day, an increasingly desperate Aldrich replied: "am in a great deal of trouble here. [General Manager J. A.] Fairleigh has proved a miserable failure & has resigned & gone. . . . I am afraid of failure & heavy loss here." Squire simultaneously sent an optimistic note: "Have no doubt about your Dudley investment turning out well

Fig. 9. Guy Gilliland and George Brewer camped in Shelby County, 1907. (Photograph by Dr. W. F. Prouty, assistant state geologist, Geological Survey of Alabama. Courtesy of the Geological Survey of Alabama Library.)

eventually. . . . Fairleigh's leaving improves your chances. . . . Keep up your spirits." Aldrich's pessimistic missive, however, passed Squire's in the mail. An obviously angry Aldrich sent a curt response to Squire: "there is no need of getting alarmed about not being over here. This Co. is about to wind up its affairs, and will therefore not need your services." Even so, Squire replied with one last encouragement: "I think better of the property than you do."[56]

Without work to do for Aldrich, Squire received timely correspondence from Eugene Smith informing him that the state legislature had increased its appropriation for geological surveys. Born in Washington, Alabama, on October 27, 1841, Smith had graduated from the University of Alabama in 1862. After enlisting in the Confederate Army and serving about a year with the Thirty-Third Alabama Infantry Regiment, he returned to Tuscaloosa as an instructor in military tactics. At war's end he left for Germany and earned his PhD from the University of Heidelberg in 1868. Back in the United States, he accepted the chair of Geology at the University of Alabama after a three-year stint as assistant state geologist of Mississippi.[57]

Smith revived publication of the *Geological Survey* in 1871, an important publication established in 1848 by Michael Tuomey but abruptly interrupted by his death in 1857. The *Survey,* commissioned by the state legislature in January 1848, followed a trend developed in other states over the last quarter century. State surveys began in the Carolinas in the early 1820s, and several states followed suit in the mid-1830s.[58] Furthermore, renowned British geologist Sir Charles Lyell visited Alabama's Warrior coal field in 1846 and commented favorably on "numerous beds of coal of good quality." Escorted by Professor Richard Trapier Brumby, chair of Chemistry, Mineralogy, and Geology at the University of Alabama, Lyell noted rich deposits of coal, iron ore, and limestone, and he predicted the industrial development of north-central Alabama. Having observed numerous outcroppings of bituminous coal, Lyell found estimated quantities impossible to determine "as no regular survey has been commenced of this State."[59]

With legislative support for renewed fieldwork, Squire contracted with Smith on May 1, 1883, to conduct a geological survey of the Cahaba coal field. Progress proved sporadic, however, because of breaks in the project caused by the lack of state appropriations. On the other hand, these interruptions afforded Squire several opportunities to pursue other personal interests. In fact, in the seven-year period from June 1, 1883, to August 12, 1890, Squire worked a total of four years and seven months on the survey with most of his efforts concentrated in the last 2.5 years. During the interim periods, he contracted with other interested parties and conducted surveys on their behalf.[60] After a quarter century of firsthand experience in the coal measures of north-central Alabama, Squire embarked on a project that would link him inextricably to the Cahaba field and lay the groundwork for its development and exploitation in the twentieth century.

3
SURVEYING AND DEVELOPING THE FIELD

1883–1910

Joseph Squire began work on the *Geological Survey* in June 1883 and continued steadily for eleven months. His work ceased in April 1884 due to Eugene Smith's budgetary constraints. Surveys resumed the following year, but were suspended again due to lack of appropriations in August 1885. Squire devoted only six months to the survey during the following two years, but he maintained a regular correspondence with Smith. In fact, most of the letters exchanged between Squire and Smith dealt with financial matters. For example, in August 1883, Smith bemoaned that he had only $300 allocated for the project for the remainder of the fiscal year. He suggested contacting Truman Aldrich to assist in funding the survey until new appropriations became effective on October 1. He lamented further: "I wish I had $10,000 instead of $5,000. Then we could go on without delay—but now the work must be done piece meal—to correspond with the money on hand."[1]

Aldrich underwrote the survey through the end of the fiscal year, agreeing to allow Squire to credit his work toward his outstanding promissory note. Due to Aldrich's loan and the influx of state appropriations in October, Squire's survey work continued uninterrupted until the spring of 1884. Focusing on the northern portion of the Cahaba field, he collected samples, examined outcroppings, and mapped his findings. In January, Smith notified Squire that only $700–$800 remained in the annual budget and asked that he collect as many samples as possible and write a preliminary report based on his current data. Specifically, he requested that Squire include the thickness of known coal seams; further information could be added in subsequent years as appropriate. In response, Squire notified Smith that he had twenty cotton bags (numbered one through twenty in indelible ink) for coal seam samples. Meanwhile, he would continue to survey the Cahaba field looking for the aggregate thickness of coal seams, the number of seams, the number of coal seams with greater than 2.5-foot thickness, and the aggregate thickness of coal in seams larger than 2.5 feet. He included his most recent calculations: 5,225-foot aggregate thickness; forty-one seams; ten seams greater than 2.5 feet; and an aggregate thickness in those ten seams of thirty two feet four inches. After an extended stay in the field, Squire returned in April with eight samples from as many seams located in the northern end of the Cahaba field. Smith instructed him to send the samples for testing to his assistant Henry McCalley in Birmingham and directed Squire to conclude his fieldwork and to begin writing his report.[2]

Correspondence continued throughout the remainder of the year, but Squire worked only one other month (October 1884) on the survey. The following spring, Smith informed Squire that he would like to have the Cahaba report and map completed and to the printers by the fall of 1886. Moreover, he wanted to include Squire's findings in his next report to the legislature. Squire agreed to have the project completed by October or November 1886, adding that he had been "making some headway lately in acquiring a better knowledge of the east part of the lower Cahaba."[3]

Work continued through the summer of 1885, when depleted budgets brought Squire's survey to a standstill once more. Things would not pick up until 1888; he worked on the survey for only six months during the two-year interim. Smith suggested that Squire seek other employment while they awaited further allocations from the state. Squire did so, informing Smith in October 1885 that he had completed two months' work for "Atlanta parties." In fact, his former acquaintance Tichenor had exhibited a renewed interest in the Cahaba coal seams.[4]

Correspondence between Squire and Tichenor from December 8, 1885, to June 12, 1888, indicates a common interest in the development of the Cahaba coal field. The technical discussion and the familiar tone indicate a bond forged through years of coal-mining pursuits. Letters written in the early part of 1886 reveal a dialogue between these two men—Squire in his office on Buck Creek in Helena, and Tichenor (writing on Home Mission Board letterhead) in his Atlanta office. Apparently, each had a detailed map of the area, and they pored over their findings to identify seams and basins. A January 23 letter from Tichenor attempted to locate the Montevallo, Thompson, Conglomerate, Gholson, and Synclinal Seams along Piney Woods and Lick Creeks. Furthermore, Tichenor included a sketch that determined a thickness of three feet or more for the Thompson Seam. He concluded that this basin offered "the finest places for mining in the Cahaba field."[5]

On June 1, Tichenor referenced Squire's diagram of the Dailey Creek area and queried, "May you not be mistaken in that cross section?" Displaying a phenomenal knowledge of practical geology, the Baptist preacher contended that the Montevallo and Dodd Seams were the same and that "the unknown vein" was 280 feet below that, rather than the 322 feet estimated by Squire. He suggested further that layers of pebbles ran above the Thompson and Montevallo Seams; that the Thompson and Gholson Seams had yellow and red sandstone pebbles; and that the Montevallo Seam lay between the Conglomerate and Helena Seams. He also inquired about the veins accessible from the basin parallel to Piney Woods Creek, as well as the outcroppings and depth of the Gholson Seam. "You see how many things trouble me," he concluded.[6]

Three days later, Tichenor excitedly wrote, "'Eureka,' 'Eureka.' I have found it. The Conglomerate vein and the Thompson are the same. There is no doubt about it." In addition, he made the following statements with certainty as he guided Squire through his evaluation of the cross sections:

> The Helena vein overlies the Thompson 126 feet and underlies the Montevallo not 322 feet but 280 feet;
>
> The Black Shale and other veins of the Buck Creek measures fall in their appropriate places in the Da[i]ley Creek series;
>
> The veins above the Montevallo are veins which must receive new names as the Helena lies below;
>
> The Thompson vein must be just under the valley of Big & Little Mayberry [Creeks] and must outcrop on both sides of the Anticlinal between the two creeks.

This evaluation hardly fits the stereotype of an otherworldly Baptist minister, but Tichenor's analysis is consistent with his interest in the development of Alabama minerals and technology as well as the acquisitive instincts of late-nineteenth-century American life. He informed Squire, "If I am right there is [*sic*] 'millions in it.' I know how close you keep a search and feel no necessity of charging you to keep this."[7]

A dialogue between the two men continued through the summer, and they exchanged theories and sketches as they refined the surveys. Tichenor implored Squire to complete the additional survey requirements and promised to buy the lands and to give Squire half interest in them. "That will be," he coaxed, "more than you would make in 40 years." They even argued from time to time. For example, on July 24, Tichenor wrote: "The more I study about it, the more strongly I think you are mistaken as to the Helena & Thompson veins on the Little & Big Mayberry." He proceeded once again to sketch his cross-sectional theories and to enumerate various thicknesses and separation distances. After dogmatically arguing his point, Tichenor concluded with a conciliatory statement: "I would like to know what you think."[8]

After visiting Squire in Helena and exploring various key parts of the Cahaba field during October 1886, Tichenor informed his friend that he had "*a big project* [emphasis in original] in view which I will communicate in due time." He did so in a letter written on November 1:

These two seams (Gholson & Spring) estimating them at 4 feet each would contain 100,000,000 tons of coal. The Thompson, Helena, Tyler, and Montevallo would contain no less than 150,000,000 more making an aggregate for the basin of 250,000,000 tons. . . . Nothing known in Alabama equals this basin.

Now for my plan—

I propose that 30,000 acres of land including about a township and a half embracing these lands and having its western branch the township line west of the Cahaba opposite the mouth of Piney Woods Creek and running eastward to the edge of the field be bought and that a city be located on the Cahaba at the mouth of Piney Woods Creek. That a railroad be built from Bucksville on the Alabama & Chattanooga through it to Calera. The Mobile and Birmingham road will run through the field crossing it. The branch of the L&N will run down to it. The Georgia Pacific will build from Leeds to it. These roads will open to us the iron ore and limestone and make "Corona," the name with which I christen our new city, the rival of Birmingham. To do this will require:

—For the lands say	$300,000
—For railroad	$250,000
—For development	$150,000
Total	$700,000

Once on its feet and the thing is worth ten times that sum. The lands above are worth today $100 per acre.

After a few brief comments, he asked, "will not DeBardeleben take hold of this for us if we demonstrate to him what is there? I know of no better man for it."[9]

In a subsequent letter dated November 11, Tichenor estimated that the proposed thirty thousand acres would contain 1.5 billion tons of coal, all within 1,500 feet of the surface. He estimated that coal operations could yield five thousand to ten thousand tons per day. In the aggregate, he projected a total value of $1.5 million (estimating one-tenth of one cent per ton). If railroads developed the area, a more accurate (although conservative) estimate would be one cent per ton, or a total value of $15 million. He concluded: "I verily believe this is the grandest enterprise in the South. If Mr. DeBardeleben will take hold of it, I am sure its success is secured." Squire replied on November 20 that he had met with DeBardeleben, had shown him Tichenor's letters of October 28 and November 1, and had made a proposal to develop the Cahaba lands. He suggested that DeBardeleben organize a company based on $200,000 capital stock to acquire the lands and to extract the coal. This endeavor would produce $10,000 each for Squire and Tichenor.[10]

From this point until June 1887, Squire and Tichenor worked earnestly to buy and sell Cahaba lands to gain control of their "Corona" area. At the same time, they contracted with DeBardeleben to provide the investment capital for development. The three agreed to meet in Birmingham on Tuesday, June 28. Squire notified Tichenor of the meeting, to which Tichenor replied: "I have just received your letter. Why are you dispirited? Give me some reason for it?" Whatever the cause of Squire's pessimism, the meeting in late June apparently made it contagious. After the meeting, Tichenor penned a note to Squire dated July 2: "I left Birmingham quite dispirited at the failure of our grand enterprise. Carried to a successful issue it ought to have made us a hundred thousand dollars apiece. As it is, the probability is that we will get nothing for our work." He then proposed that Squire persuade Aldrich or some other investor to buy DeBardeleben's interest (about $30,000) with the intent of investing an additional $60,000: "This is the biggest thing in Alabama, and I dislike to lose it."[11]

DeBardeleben did ultimately sell the lands to Aldrich. In December 1887, Squire informed Tichenor about the deal and that their profits would be $5,000 each. Obviously frustrated, Tichenor replied, "If DeB[ardeleben] could have carried out his contract with us our profits would have been $50,000 each instead of $5,000. As we released him from the obligation to complete the contract, I think we ought to have all the profits." Unfortunately for him, DeBardeleben did not agree. In fact, title disputes delayed closure on the deal until June 1889. By that time, Squire was hard at work on the *Geological Survey;* Tichenor had shifted his focus to iron ore; and each man received only $5,000. Disappointed, Tichenor complained, "It has cost us 20 years of labor."[12]

With the demise of their grand scheme, Squire and Tichenor corresponded less frequently. Aldrich continued to pepper Squire with inquiries—questions about land titles and requests for land purchases, maps, and information about coal seams. Despite such disruptions, Squire turned his attention to the *Geological Survey* in February 1888. This project would consume his efforts for the next 2.5 years until the map was in the hands of the engravers and the text had been sent to the printers. Smith asked Squire to complete the map first so engravers could be working on it while he finished the text. Furthermore, he wanted to send the report to the printers that summer and present the final product to the legislature in November. Squire could not respond that quickly, but he did continue to work diligently on the project. On the other hand, he regretted having to turn down other, more lucrative offers for survey work, informing Tichenor that he would "feel like the day of jubilee has come when I get it completed."[13]

A similar exchange between Smith and Squire occurred the following spring. Smith encouraged Squire to complete the map so it could be sent to the engravers during the summer months, their period of reduced workload. In turn, Squire promised to finish his tracings by July, but requested a typewriter clerk and a draftsman to assist him. Smith agreed to hiring the assistants, but noted that all payments might be delayed until October 1 when new budget appropriations would be available. Fortunately, funding continued, and Squire persisted in his work.[14]

The final year of drafting and writing assumed the harried nature of most projects as they approach deadlines. In October, Smith informed Squire that the map should bear the title "Geological Survey of Alabama" rather than "State Geological Survey of Alabama." In addition, he suggested that the legend read: "Map of the Cahaba Coal Field and adjacent geological formations: constructed (or compiled) from notes of numerous geological surveys made (by Joseph Squire) [optional] during the past thirty years. Surveyed and drawn by Joseph Squire, Mining Engineer." In April, Smith asked Squire: "have you got your 'sign' hung up—reminding you that the time for going to press is not far off?" Squire replied that his sign, consisting of letters two inches high, reminded him to "Get Cahaba Field Report finished by July 1, 1890." A few days later, Squire left to collect coal samples from the Henry Ellen, Montevallo, and Brierfield Seams. Upon his return, he informed Smith that he had received four samples of coal and three of coke from Blocton and that he had requested samples from the Gholson, Helena, and Wadsworth Seams. Promising to forward the samples to Smith when the others arrived, Squire returned to his drafting. Smith responded: "I am appalled when I think of the amount of work there is to do and the little time there is to do it in." Nevertheless, within a month, Smith and Squire met to discuss minor changes to the final proof of the Cahaba map before sending it to the engravers.[15]

In addition to mapping the coal seams and basins on a topographic map, Squire included detailed marginal data. For example, to provide comprehensive, three-dimensional information, he developed cross-section graphs of each seam. To complement the numerous cross sections, Squire constructed thirty-six diagrams for inclusion in the publication. Even so, as the July 1 deadline approached, Smith felt compelled to prompt Squire again: "Have you got it about ready?" Squire had not. Once again he filled his letter with a litany of excuses: he had been sick for a week; he had burned his left hand; he needed more clerical support. Once those problems were solved, he promised, he could finish the report in two weeks. Various prompts, explanations, entreaties, and suggestions continued for more than a month until Squire shipped maps, sections, specimens, and plats to Smith. On August 12, 1890, he recorded in obvious relief that he finally "brought my Cahaba Coal Field work to a finish."[16]

Over the course of three decades, Joseph Squire established himself as the ultimate authority on the Cahaba coal field. In his private enterprises, he served as miner, prospector, geologist, operator, and discoverer. Ethel Armes stated that "there is no coal or iron man of Alabama to whom the name . . . of Joseph Squire [is] unknown." Squire depicted himself as a geologist and mining engineer, specializing in coal and iron ore lands and mines. By his own admission, he "had the longest experience in examining the coal lands and superintendence of mines of the Cahaba and Warrior coal fields of any one in the state." He developed a list of nearly 30,000 acres of Cahaba coal lands identified by range, township, and section. He knew which sections contained coal, which did not, which were vacant, and who owned the surface and mineral rights. Although the wealth he sought from this work proved elusive, his "Report on the Cahaba Coal Field" in the 1890 *Geological Survey of Alabama* constitutes a timeless work. As Smith predicted to Squire in November 1888, "your report will bring you more credit and ultimately 'finances' than any other work undertaken."[17]

After completing the 1890 *Survey,* Squire spent the next few years drawing field maps and condensing old field notes. Even though he continued some field explorations, he declined numerous offers for work in the Cahaba field. He maintained contact with Smith, evaluating soil samples and reviewing a report on coal seams in St. Clair and Blount Counties. In the fall of 1893, Squire spent two weeks in the Blount Mountain area, returning with firsthand knowledge and a roughly sketched map of those coal lands. He determined that "Mr. Gibson's report is generally speaking correct," and Smith proceeded to publish the study on Squire's recommendation. Due to publication costs, Smith wanted to publish the report without an accompanying map, but Squire deemed the map essential. He noted: "The coal field is a wild broken country and out of the line of travel, and portions of the field will most certainly be developed in the near future."

Such development would have to come from sources other than Squire. Two entrepreneurs stand out in the development of the Cahaba coal field. As Squire admitted: "It was not I who brought the capital into Alabama! No, it was not I! I left that to Aldrich and DeBardeleben."[18]

Truman Heminway Aldrich left his banking business in Selma for the Cahaba coal field in February 1873. He would become one of the foremost developers of the mineral resources of central Alabama, but his entry into the coal business has direct links to the pioneering efforts of William Phineas Browne and the Alabama Coal Mining Company (ACMC). Writing in 1875, Aldrich described the sequence of events that led to his eventual purchase of the Irish Pit and surrounding coal lands. He noted that the Cahaba Coal Company (CCC), the enterprise that replaced William Phineas Browne as agent for the Mobile and Selma Coal Mining Company, operated the Montevallo mines for approximately three years. After extending the main gangway to a distance of one thousand feet, officials leased the mines to D. A. Smith in 1871. When Smith's lease terminated two years later, Aldrich bought the entire interest and began his coal-mining pursuits.[19]

Meanwhile, the Montevallo Coal Mining Company, successor to the ACMC, leased its property to E. G. Walker until it finalized the sale of its mines and coal lands to the Central Mining and Manufacturing Company (CMMC). Establishing its operations within close proximity to the Irish Pit, the CMMC opened a six-hundred-foot slope and began long-wall mining in 1867. Josiah Morris and others mined much of the CMMC coal until the CCC negotiated a twelve-month lease in 1870. For a brief interval, the CCC established unilateral control over all of the Montevallo mines, but this virtual monopoly was short-lived. In 1872, Albert Williams, S. D. Holt, and other investors formed Holt, Varner, and Company and assumed control of the old ACMC mines. An 1874 newspaper report identified two mining operations in the Cahaba field: Holt's machine-operated mine that produced sixty to seventy tons per day, and the ACMC drift mine with a daily output of eighty tons. The account stated that "these two mines furnish nearly *three fourths* of all the Alabama coal used in the State." These mines were abandoned by the end of the year, although T. S. Alviss and Company mined a small amount of coal from the Lemley vein, converted the coal to coke, mixed the coke with charcoal, and used the fuel in the puddling furnace at Brierfield.[20]

By the mid-1870s, Alabama's economy seemed to have recovered from the devastation of the Civil War. W. David Lewis, chronicler of the history of Sloss Furnaces, reports that Alabama became the South's leading producer of coal and iron by 1875. Wayne Cline, focusing on the development of Alabama railroads, notes that "after the Civil War, coal would supplant wood as locomotive fuel, and abundant Alabama coal would provide energy for steam engines around the world." Furthermore, Justin Fuller asserts that boomtowns emerged as an integral part of the growing iron industry and its development of blast furnaces. Certainly, the Birmingham District, spawned in 1871, would lead the way in central Alabama's industrial development, but raw materials such as coal, iron ore, and limestone formed the foundation for such growth.[21]

In addition to the Montevallo mines, other enterprises began to dot the landscape of the Cahaba coal field during the postwar period. For example, the CCC got its start in 1866 when it bought the Woodson & Gould mine near Helena. This is the same organization that would lease the former Browne mines in 1868 and the former ACMC mines in 1870. After selling to the CCC, William Gould and other investors formed the Glasgow Coal Company and opened a new mine for producing coke. Gould's enterprise was discontinued for a time, but D. C. Bozeman was mining coal for blacksmiths by 1875. The Alabama Mining and Manufacturing Company continued the work begun by F. L. Wadsworth who opened the Eureka mine near Helena in 1867. The Red Mountain Iron and Coal Company conducted operations from 1868 to 1872, and a Mr. Leavett and others leased the mine until the Alabama Mining and Manufacturing Company took charge in 1874. Operators sank a new slope at Davis and Carr's mine on the Cahaba Seam in 1873, and Byram and Bowers mined two points on the Helena Seam formerly operated by the Red Mountain Iron and Coal Company. Independent operators tapped the Shortridge Seam in 1874, and D. C. Bozeman opened a slope on the Little Pittsburgh Seam the following year. Thus, in spite of the Panic of 1873, the Cahaba coal field offered numerous opportunities for individuals and small companies alike. Furthermore, the emergence of these relatively minor operations suggests that some southern industrialization occurred during Reconstruction and prior to the New South period (1880–1900).[22]

Still, Truman Aldrich represented the primary developer of Cahaba coal. But he left his Montevallo interests for the Birmingham District in 1877. This move resulted directly from Henry Fairchild DeBardeleben's entry into the Cahaba field. As mentioned earlier, DeBardeleben and Daniel Pratt were involved in the purchase and rebuilding of the furnaces at Oxmoor (1872–1873). This enterprise, initiated by Frank Gilmer and John T. Milner a decade earlier, had remained dormant since its destruction by Wilson's Raiders in 1865. As local resident Mary Gordon Duffee described, Oxmoor portrayed "a scene of loneliness and ruin that makes my soul faint to recall it." In spite of failing health, Pratt agreed to finance the restoration project, an endeavor that Pratt biographer S. F. H. Tarrant considered "the last and crowning act of his life." Purchasing several abandoned mines, the two damaged blast furnaces, and approximately 6,500 acres of iron ore lands on Red Mountain, Pratt and DeBardeleben gained a majority interest in the Red Mountain Iron and Coal Company. They changed the name to the Eureka Mining Company and selected Judge Henry D. Clayton of Eufaula as the new president. DeBardeleben assumed duties as superintendent and general manager, but success did not follow immediately. Clayton and DeBardeleben struggled to minimize production costs, generate markets, employ skilled labor, and maintain a sufficient supply of timber for converting to charcoal. Admitting that this opportunity represented his "first lesson in the iron business," DeBardeleben commented later that "I came in and took charge of what I knew nothing about!" Frustrated, DeBardeleben resigned his position and forfeited his salary for the previous year. Shutting down the furnaces, he returned to Prattville to pursue more familiar interests.[23]

A new syndicate formed in the fall of 1873, and the Eureka Mining and Transportation Company of Alabama (EMTC) absorbed the rights and titles of both the Eureka Mining and the Red Mountain Iron and Coal Companies. Daniel S. Troy served as president, and Levin S. Goodrich assumed DeBardeleben's former duties as manager and superintendent. Putting the refurbished furnace into blast the following winter, the EMTC experienced many of the same hardships endured by Clayton and DeBardeleben. Goodrich had increased efficiency by reducing fuel consumption and expanding output, but his systematic approach to production suggested that coke would constitute a better fuel. Unable to devote capital to such experimentation, the EMTC struggled in vain to produce sufficient quantities of charcoal iron.[24]

Facing an ever-increasing debt of $240,000, EMTC officials made a public offer to relinquish control of the Oxmoor furnaces "to any man or any company of men desirous of proving that iron could be successfully manufactured in the Birmingham District." Convening a meeting of capitalists interested in the development of central Alabama, John T. Milner proposed the formation of yet another company to provide additional capital to facilitate the tests advocated by Goodrich. Ultimately, Milner and his associates formally organized the Cooperative Experimental Coke and Iron Company and elected Goodrich as superintendent. James W. Sloss, Charles Linn, and William S. Mudd would serve as a board of managers.[25]

Goodrich began immediately to convert the furnaces from charcoal to coke production, and Frank P. O'Brien constructed five coke ovens designed by a Belgian inventor named Shantle. At this point, several mining operators supplied samples of their coal for Goodrich's coking experiments. In the final analysis, Goodrich determined that coal from the Browne Seam of the Warrior coal field was the most suitable coking coal. This seam, discovered by "Uncle Billy" Gould and named for William Phineas Browne, represented the salvation of iron production in the Birmingham District. Renamed the Pratt Seam in honor of Daniel Pratt, these coal deposits would become the lifeblood for Birmingham's iron industry. The Oxmoor experiments culminated in the production of coke iron on March 11, 1876.[26]

Meanwhile, Truman Aldrich, with Joseph Squire's assistance, had investigated coal lands in both the Cahaba and Warrior fields in an attempt to augment his Montevallo mines. As he explored this region, Aldrich determined to find coal suitable for large-scale iron production. He based his search on three criteria: locating a first-class coking coal embedded in seams thick enough to be extracted easily and economically; identifying sufficient quantities of the coal to warrant considerable investment of capital and time; determining the existence of such coal within close proximity to Oxmoor, thereby making it available for iron production and for railroad use. He, too, found the Browne Seam and, with Squire's help, purchased most of the property containing the coal veins. Teaming with James W. Sloss, Aldrich raised $30,000 and established mining operations in the Warrior coal field.[27]

Henry F. DeBardeleben reentered the picture at this point as well. Having

Fig. 10. William "Uncle Billy" Gould, coal-mining enthusiast and entrepreneur. (Courtesy of Hugh Gould.)

inherited Daniel Pratt's estate upon the death of Pratt's wife, Esther, in 1874, DeBardeleben acquired a controlling interest in Oxmoor stock. Selling his interest in the Eureka Company to Sloss for $160,000, DeBardeleben provided additional capital for Aldrich's pursuits. He joined Aldrich and Sloss in January 1878 to form the Pratt Coal and Coke Company. DeBardeleben became president, Sloss served as secretary and treasurer, and Aldrich assumed duties as superintendent and mine manager. Opening several slopes and building the Birmingham & Pratt Mines Railroad, the triumvirate quickly established a direct link, both economically and geographically, between their supply of coking coal and the furnaces of the Birmingham District. These captive mines would feed the Birmingham iron industry and eventually spawn the development of the mining community known as Pratt City. With the iron industry well in hand, these coal developers forsook their mining roots and turned toward the promise of wealth in the Warrior field. Thus, the ultimate success of the Oxmoor experiments spelled doom, at least for a time, for the Cahaba coal field.[28]

As the southern regions lay dormant, Jones Valley, located just six miles to the north, boomed as Birmingham earned its reputation as the "Magic City." Having invested $125,000 in the Oxmoor experiment, M. H. Smith of the Louisville & Nashville Railroad (L&N) committed himself to completing the north-south link initiated by John T. Milner, Frank Gilmer, and the South & North Alabama Railroad nearly two decades earlier. That patchwork line had served the Cahaba field's fledgling mining operations, but the promise of exponential increases in iron and coal production called for more efficient transportation. In fact, coal production increased to 97,000 tons within a year of the Oxmoor experiment. Milner stated that "this success is due mainly to the great effort now being made by the railroad management [i.e., Smith] to forward and develop this business." Likewise, Ethel Armes concluded that Smith's cooperation and encouragement proved critical in the practical development of the Birmingham District. According to Armes, "his hand, back of the Louisville and Nashville Railroad management, was the great shaping force."[29]

As the Pratt Coal and Coke Company established itself as the largest mining operation in the Birmingham area, each of its primary officials assumed a distinct role. DeBardeleben, ever the great "horse trader," served as promoter. Having secured capital, land, and labor to ensure steady production, he traveled the countryside finding buyers, establishing markets, and developing a consistent demand for Pratt coal. Aldrich, on the other hand, remained at the mining camp. Charged with sustaining production, he managed mine operations, coordinated transportation assets, supervised miners, and maintained machinery and equipment. In addition, Aldrich conducted necessary surveys, implemented engineering projects, and pursued his interest in "geologizing." Sloss assisted in coal production initially, but left the Pratt Company in late 1879 to manage the Oxmoor Furnace. In that capacity, he addressed both technological and financial difficulties, but his ultimate frustration derived from conflicts with stockholders based in Louisville and Cincinnati. After struggling for two years to meet rising demands, Sloss complained to DeBardeleben, and the latter replied: "Why don't you build you a couple of furnaces of your own, man? I'll let you have my Pratt coal at cost, plus ten percent for five years." Backed by DeBardeleben, Sloss contracted with Mark W. Potter for red iron ore, with L&N president, E. D. Standiford, for transportation assets, and with B. F. Guthrie for additional capital. Consequently, he terminated his interest in the Eureka Company and founded the Sloss Furnace Company in 1881.[30]

Aldrich also left the Pratt Company and established his own Cahaba Coal Mining Company (CCMC) in 1881. According to historian Justin Fuller, DeBardeleben tended to manage by domination and, therefore, did not work well with associates. Aldrich's resignation left DeBardeleben with sole ownership of "the largest and most successful coal and coke company in Alabama." That distinction would prove short-lived as Aldrich built a colossal enterprise of his own. First, Aldrich sold his Henry Ellen properties—named for DeBardeleben and his wife, Ellen Pratt DeBardeleben—to his former partner for $400,000. Then, combining more than twelve thousand acres of coal lands stretching through Jefferson, Shelby, and Bibb Counties with $1 million in capital, Aldrich organized the CCMC. He served as president and treasurer, Cornelius Cadle became vice president and general manager, Peter B. Thomas superintended the mines, and Lewis Minor of Connellsville, Pennsylvania, was general superintendent responsible for coke production. Additionally, Aldrich and Cadle served on the board of directors along with W. S. Gurnee of New York and Samuel Noble and A. L. Tyler of Anniston. (Noble and Tyler were northern industrialists who had developed Anniston as a company-owned iron town.) Ultimately, CCMC properties would comprise more than 1.2 million acres of coal lands.[31]

Based at Blocton in north Bibb County, the CCMC first established a railroad connection to its mines. A line extending from Woodstock in 1883 constituted the Woodstock & Blocton Railway and provided a connection with the Alabama Great Southern Railroad. With access to coal and coke markets, the CCMC mined and shipped its first coal in the spring of 1884. Oxmoor and Anniston became the primary markets for coke, and the CCMC delivered steam coal to all but one of the railroads operating out of New Orleans. With these expansive markets, the CCMC surpassed DeBardeleben's Pratt Company as the largest supplier of coke and coal.[32]

Aldrich's success in the New Orleans market may be attributed to A. C. Danner, one of his associates in mining operations at Dudley in 1882. According to Danner, Alabama mine operators endured the Panic of 1873 and produced a surplus of coal the following year. Coincidentally, low water in the Kanawha and Ohio Rivers preempted shipments from Pittsburgh, the traditional source of coal for New Orleans. Also, a severe coastal storm wrecked and sank numerous barges loaded with Pittsburgh coal. Seizing the opportunity and receiving cooperation, encouragement, and assistance from M. H. Smith of the L&N Railroad, Danner arranged to ship Alabama coal directly from the mines to New Orleans at a rate of $2.25 per ton. Danner negotiated a contract with the gas company for twenty thousand tons of Alabama coal, but he encountered considerable prejudice among New Orleans engineers. John T. Milner traveled to New York and convinced officials of the Southern Pacific Railroad Company to test Alabama coal in their locomotives. Ignoring threats of strikes by their engineers, company officials contended that steam coal used on Alabama, Georgia, and Mississippi railroads would be sufficient to power locomotives for the Southern Pacific. The test proved successful, and the Southern Pacific soon consumed five hundred tons per day in its run between the Crescent City and San Antonio.[33]

Having established a steady market for coal in New Orleans, Aldrich expanded his purview. An additional railroad line—the Brierfield, Blocton, & Birmingham—completed in 1890 and running from Gurnee Junction to Bessemer, provided further access to growing markets. Consequently, Aldrich organized two new companies—the Excelsior Coal Mine Company and the Export Coal Company (ECC)—and opened two mines in Shelby County near Gurnee Junction. Eventually, the Excelsior Company would be absorbed into the CCMC, but the ECC developed its own niche market. With W. D. Munson of New York as president, Aldrich as vice president, and M. P. Canfield as secretary and treasurer, the ECC pioneered coal exports to the West Indies, Mexico, and numerous points in Central and South America. Combining assets—a steamship line from Munson, operating capital from the CCMC, and $100,000 from M. H. Smith and the L&N Railroad for constructing a wharf at Pensacola—the ECC enjoyed a virtual monopoly of the coal markets to the south. As Aldrich commented, "the natural outlet of the Cahaba Company was to the South. . . . The export business grew out of the necessities for new markets." Once established, the ECC owned and operated two steamships, two schooners, six barges, and two ocean-going tugboats. Altogether, Aldrich's mining interests comprised "the best and largest area of good coal land ever gotten together in the State up to that time."[34]

Things were not going as well for Henry DeBardeleben. After Sloss's and Aldrich's departures from the Pratt Company, he began to experience health problems. Fearing that he had tuberculosis, he determined to sell his Alabama interests and move to the drier climate of northern Mexico. He found a buyer in Enoch Ensley, a member of a syndicate of financiers based in Memphis. Ensley, who had attempted unsuccessfully to buy stock in the Nashville- and Chattanooga-based Tennessee Coal, Iron and Railroad Company (TCI), sought to settle a grudge match. Attempting to introduce west-Tennessee capital into the Birmingham District before east-Tennessee interests moved southward, Ensley's group agreed to pay $600,000 over six years and gained control of the Pratt Company on December 29, 1881. Signing over his other mineral interests to Alice Furnace associate F. L. Wadsworth as trustee, DeBardeleben commenced sheep ranching near Laredo, Texas. Gaining strength in the open air and learning that he did not have tuberculosis, DeBardeleben made periodic trips to San Antonio to mingle with capitalists seeking to develop regions of the southwest. In October 1881, he met William Thompson Underwood, a lawyer from Kentucky who investigated faulty land titles. Hearing DeBardeleben's stories of real estate ventures and industrial enterprises in central Alabama whetted Underwood's appetite for speculation. Both men made their way to Birmingham and formed the DeBardeleben and Underwood Company in February 1882. Buying thirty acres of land and building the Mary Pratt furnace, they began operations in 1883. DeBardeleben attempted to develop his Henry Ellen properties and prospect in the Blue Creek Basin, but recurring health problems forced him to return to his haven in Mexico. Consequently, furnace operations devolved to Underwood.[35]

Regaining his health once more, DeBardeleben returned to Birmingham in

1885 to "make smoke." This time, he established a partnership with David Roberts, an English banker who immigrated to the United States and married into a prominent Charleston, South Carolina, family. Inspired by DeBardeleben's accounts of Birmingham's development, Roberts joined DeBardeleben in forming an enterprise to develop nearly thirty thousand acres of the latter's lands. As the partners made final arrangements, DeBardeleben boasted: "Here [pointing to a spot on a map twelve miles southwest of Birmingham] is where I am going to establish a young city. . . . We are going to build a city that will contain eight furnaces within two years. . . . We are going to build a city solid from the bottom up and establish it on a rock financial basis. No stockholder will be allowed to come in who can't make smoke. It will take $100,000 to come in, and the man who can make the most smoke can have the most stock." DeBardeleben received 50 percent of the stock in exchange for his lands, and Roberts gained 5 percent based on his efforts in raising additional capital. The remaining stock would be distributed to English and South Carolina investors recruited by Roberts.[36]

DeBardeleben and Roberts incorporated two companies in March 1886. The DeBardeleben Coal and Iron Company (DCIC) would develop ore and coal mines and build coke ovens and blast furnaces. DeBardeleben became president, Roberts was vice president and general manager, and Andrew Adger served as secretary and treasurer. The Bessemer Land and Improvement Company focused on land speculation, selling its "boom-town" lots to anyone interested in developing a new city named for English steel master, Sir Henry Bessemer. As an incentive, DeBardeleben offered a 5 percent discount on coal, coke, and iron to industrialists willing to build plants in Bessemer. Both companies proved successful, and DeBardeleben merged them with his Henry Ellen Coal Company and the newly acquired Eureka Company in 1889. Following this consolidation, the enlarged DCIC claimed capitalization of $13 million, an astronomical figure for Birmingham industries at that time. Basking in his success, DeBardeleben quipped: "And every sheaf in the field rose up and bowed to my sheaf!"[37]

At the same time that DeBardeleben and Roberts built the city of Bessemer to rival Birmingham's industrial center, the national economy made a dramatic upturn. Shortly thereafter, TCI—Ensley's nemesis—"burst like a meteor upon north Alabama late in 1886 as the boom surged irresistibly forward." TCI's expansion began in 1881 as John Inman directed the absorption of iron-making resources in east Tennessee. Inman, a Tennessee native, moved to New York after the Civil War and established himself as a shrewd businessman. Adhering to Henry Grady's depiction of a New South, he invested regularly in southern industries. On the other hand, Justin Fuller describes Inman as "one of the most unprincipled manipulators of securities to operate in the South." Rather than assisting companies in which he invested, Inman frequently precipitated their ruin due to his speculative approach to business. Nevertheless, under his direction, TCI became one of the largest coal and iron enterprises in the nation.[38]

TCI's entry into Birmingham came as a result of its purchase of Ensley's Pratt Coal and Iron Company. With this merger, TCI gained the Pratt mines that produced more than three thousand tons of coal daily, two furnaces (and four under construction), 1,210 coke ovens (with 1,800 under construction), 76,056 acres of coal lands, and 12,204 acres of ore properties. In sum, TCI emerged as the largest producer of coal and iron in the South. Once established in the Birmingham District, Inman began his trademark maneuvers that earned the company a reputation as a "football for speculative gambling." Having ridden the national upsurge through the early months of 1886, TCI stock had increased in value from $15 to $40 per share. By September, stock prices had risen another $30, but Inman sought further increases. By resorting to real estate speculation and an increase of $1.5 million in company stock, Inman coaxed the price of TCI stock to an all-time high of $118 by the end of the year.[39]

Inman's success was short-lived. Due to the speculative nature of the increased value, TCI stock was forced to ride the peaks and valleys of the volatile economy. Furthermore, land speculation dictated that, because the original purchasers had no intention of residing in the area, real estate prices were based on projections of future sales rather than on current value. Consequently, as the boom of 1886 cooled, TCI stock plummeted to $21 by September 1887. Ultimately, Inman's strategy precipitated his undoing. Having based the company's growth on a shaky foundation, he was replaced by Thomas C. Platt in 1889. Platt proved no different in his approach to stock manipulation. As historian W. David Lewis recounts, Platt manipulated TCI securities as ruthlessly as Inman and spawned a financial panic that rivaled that of his predecessor. Also, the national economy became sluggish by 1890. This downturn resulted from a scarcity of capital, competition from other booming areas, and the predictable loss of momentum inherent in any speculative venture. As a result, Inman regained control of TCI in 1891 and set the stage for yet another phase of expansion.[40]

Early in 1892, TCI officials met in New York to discuss recent decreases in iron prices. Designing a strategy to consolidate production, they extended invitations to their Birmingham competitors to join the parley. TCI offered to purchase the Sloss Iron and Steel Company and the DCIC, but Sloss's representatives balked at the offer. In contrast, DeBardeleben welcomed the merger, and the two companies signed an initial agreement on March 18. TCI would exchange $8 million of TCI stock for $10 million of DCIC stock in addition to assuming responsibility for an additional $3 million of bonded indebtedness. Stockholders of both companies ratified the agreement, and the consolidation took effect on June 1.[41]

In addition to the Sloss and DeBardeleben companies, three smaller enterprises participated in meetings with TCI. The Woodward Iron Company, the Pioneer Mining and Manufacturing Company, and the CCMC all received offers for consolidation. Truman Aldrich, founder of the CCMC, was the only participant to agree to a merger. In September 1892, TCI absorbed the CCMC, exchanging $3 million in stock for the latter's coal and ore lands.[42]

The mergers of 1892 doubled the size of TCI. Capital stock increased from $10 million to $21 million, and land holdings expanded from 210,000 to 400,000 acres. Daily coal production surged from 7,000 to 13,000 tons, and seventeen blast furnaces produced 633,000 tons of pig iron annually. With its increased assets for pig iron production, TCI ranked third nationally behind the Illinois Steel Company and the Carnegie Company. The new board of directors ousted Platt from the inner circle and reinstated Nat Baxter as president and James Bowron as secretary-treasurer. In addition, officials approved the appointment of Henry DeBardeleben and Truman Aldrich as first and second vice presidents, respectively. Other officials included David Roberts, third vice president; G. B. McCormack, assistant general manager; and Erskine Ramsay, chief engineer. Even though DeBardeleben owned a majority of TCI stock, he held insufficient resources to dominate management. Consequently, he struck an agreement with Inman to share direction of the company.[43]

In the wake of TCI's consolidation, the national economy experienced another nosedive. The Panic of 1893 ushered in four years of economic depression. Even so, DeBardeleben, seemingly taking his lead from Inman's former exploits, sought to take advantage of dramatic shifts in Wall Street. Expecting a rebound in the market, DeBardeleben formulated a plan to "bull" TCI's stock. In reality, much of his stock was held on margin, his 120,000 shares were insufficient to manipulate stock values, and the national economy continued in its downward spiral. Furthermore, TCI's speculative past brought it to the brink of bankruptcy in the early 1890s. From February to July 1893, TCI stock declined from $34 to $10 per share. At the end of March, DeBardeleben sold 25,000 shares to Inman in an attempt to generate enough cash to keep him afloat. The bear market continued, however, and DeBardeleben transferred the remainder of his stock to Inman's control by the end of May. Having lost his fortune, estimated at $2.5 million, in a matter of weeks, DeBardeleben commented: "I met my Waterloo. . . . They cut my claws, and you can't fight without claws. When my money went I came to hear men's voices change, see their expression change. . . . And I clawed the rocks."[44]

In spite of his losses, DeBardeleben remained with TCI as first vice president. He continued to wield influence, but his lack of economic clout compromised his position. Consequently, Inman and his New York directors moved to oust DeBardeleben in 1894. After gaining assurances that TCI would contract for iron ore with his new mining enterprise, DeBardeleben resigned on October 22. Three days later, TCI's board of directors repudiated the ore contract. As Fuller reports: "On this ignominious note Henry DeBardeleben's career with the Tennessee Company came to a close." DeBardeleben, down but not out, continued in the coal and iron business. Organizing the Alabama Fuel and Iron Company, he moved into St. Clair County and opened two coal mines at Acmar.[45]

Aldrich, too, resigned from TCI in 1894, ostensibly over disagreements with DeBardeleben. Later that year, he accepted the Republican nomination (with Populist endorsement) for Congress from the Birmingham District. Losing in a close vote to Democratic candidate Oscar W. Underwood, Aldrich contested the election results. Eventually winning a judgment in his favor, he gained the congressional seat in June 1896. This settlement constituted a pyrrhic victory of sorts inasmuch as the term of service ended in March 1897. Forsaking his political interests at that juncture, Aldrich returned to the coal and iron industry of central Alabama.[46]

Aldrich negotiated a contract with Joseph Squire shortly after leaving TCI "for the purpose of exploring and acquiring, testing, developing and selling the mineral lands containing gold, silver or other metals in the part of Alabama known as the Metamorphic region." He organized the Cahaba Southern Mining Company at Hargrove in Bibb County in 1896 and the Brilliant Mining Company at Winfield in 1898, the latter constituting a partnership with his son, Herbert. In 1900, Aldrich assumed duties as second vice president and manager of mines for the Sloss Iron and Steel Company, a subsidiary of the Sloss-Sheffield Steel & Iron Company. Leaving Sloss in 1901, he organized a number of mining ventures before repurchasing his former coal lands near Montevallo. Becoming president of the Montevallo Coal Company in 1905, he admitted that mining operations were improving but that an unusually warm winter had made coal extraction "rather slow business." By January 1906, Aldrich was in Tallapoosa County mining for gold at Hog Mountain and Silver Hill. The Hillabee Gold Mining Company, based in Alexander City, coordinated prospecting ventures between that city and Dadeville.[47]

Truman Aldrich remained involved in industrial development, scientific research, and geological study until his death in 1932. He outlived his primary associates and, therefore, gained some historical perspective on their exploits during the late-nineteenth century. James W. Sloss died in May 1890, thereby missing many of the key decisions faced by his company—rejection of TCI's consolidation offer, preservation of a developmental approach to industry, and determination to continue producing iron when TCI converted to steel production. Henry F. DeBardeleben left his mark as well, even after losing his fortune and being ousted from TCI. Justin Fuller concluded that "sadly for DeBardeleben the boom times for the individualistic entrepreneur in the Birmingham district were over." In one sense,

Fuller is correct. On the other hand, both DeBardeleben and Aldrich ultimately returned to their roots in the Cahaba coal field. Their entrepreneurial endeavors brought new life to the Cahaba field and set the stage for the construction of numerous coal towns throughout the region.[48]

In spite of Aldrich's relative longevity, DeBardeleben's death in December 1910 and Joseph Squire's demise in October of the following year marked the end of an era. Building on the pioneer work of William Phineas Browne, the quadrumvirate of Aldrich, DeBardeleben, Sloss, and Squire explored, sampled, charted, mapped, exploited, and developed the Cahaba coal field. Mining efforts in the last half of the nineteenth century laid the groundwork for the golden age of coal in the first half of the twentieth century. As Lewis explains, the national economy gained momentum from its recovery in 1897 through the following decade. The first ten years of the twentieth century marked a doubling of U.S. foreign trade and a trebling of deposits in national banks. In addition, circulating currency expanded by $1.2 billion, and the country's financial institutions increased the value of their assets to exceed $21 billion.[49]

Worldwide and national prosperity would be manifested in the Cahaba coal field as well. As the iron and steel industry of the Birmingham District tapped the coal reserves of the Warrior field, smaller enterprises developed in the Cahaba valley. From St. Clair County to Bibb County, from Margaret to Lucile, numerous capitalists tapped the veins of Cahaba coal. Building on nineteenth-century foundations, entrepreneurs sought their fortunes among the "black diamonds" of Cahaba. Once established, these coal-mining ventures attracted waves of workers and their families. Communities developed, and coal towns emerged. With this migration from the farms to the mines, people brought life to the Cahaba coal field. In contrast to the chronological record of key developers in the nineteenth century, the thematic account that follows of the twentieth-century history of Cahaba shifts focus to the ordinary people who sustained the industry—their joys, their sorrows, their lives.

4
COAL TOWNS

1881–1919

One of the first bona fide coal towns of the Cahaba field developed at Blocton in north Bibb County in 1881. Established as the home of Truman Aldrich's Cahaba Coal Mining Company (CCMC) in 1883, "Blockton" opened its first post office in March of the following year. By the fall of 1884, the CCMC had constructed one hundred houses and projected another one hundred dwellings for the near future. At that point, Blocton contained a union church, a schoolhouse, an Odd Fellows Hall, and approximately a thousand residents, thereby gaining distinction as the county's largest community.[1]

Situated among a series of hills and valleys known as the Cahaba Ridges, Blocton faced numerous geographical challenges in its early years. Dubbing the rugged terrain as "The Uglies," four hundred workers bridged Caffee Creek ten times in order to establish a rail link with nearby Woodstock. Completed in January 1884, this eight-mile connection with the Alabama Great Southern Railroad enhanced mining operations immediately. Other rail links followed. The Brierfield, Blocton, & Birmingham Railroad (the "Old BB&B") connected Blocton to points eastward after completion of its Cahaba River trestle in 1890. By 1898, the Georgia, Mobile, & Ohio Railroad ran northward from Eoline to Blocton and facilitated other mining operations at Moffat, Red Eagle, and Lucile.[2]

As its transportation network improved, Blocton experienced considerable growth. By 1890, the town contained more than four hundred family dwellings and company structures, and its population swelled to 3,600. Success built on success, and, as noted previously, Aldrich sold the CCMC and his Blocton mines to the Tennessee Coal, Iron, and Railroad Company (TCI) in 1892. Local historian Charles Adams contends that this merger with TCI marked "the end of an era in the young community's history," but growth continued even with the change in company leadership. Within two years of the TCI acquisition, Blocton's population exceeded five thousand, twenty trains serviced the community daily, and the mines produced more than five thousand tons of coal per day.[3]

Ultimately, Blocton mining operations encompassed ten distinct mines. Opened for operations in 1883, each of two slope mines employed a 100-horsepower engine to hoist coal cars from the mines to the tipple. A wire rope or cable worked around a series of spindles to move "trips" (consisting of five to ten cars) along the slope. At Number 1 Mine, the railroad yards, depot, and tipple occupied a level area near the mine entrance. Here, top-side workers weighed, recorded, and dumped coal cars to fill rail cars positioned below the tipple. A double track extended the full 875-foot length of the slope, thereby providing for coal trips and the less frequent "mantrips." A ventilation slope ran parallel to the "manway," and a ten-foot diameter fan circulated fresh air throughout the mine. In addition, a steam engine pumped water from the mine floor. At Number 1 Mine, miners worked a seam that included four distinct veins of coal.[4]

Most of these miners lived in Number 1 Town, a settlement connected to the mine entrance by a footpath. In addition to their homes, the community included Baptist, Methodist, and Presbyterian churches; a school; and homes for the preachers and doctors. Company offices, the mule lot, and the harness shop lined the footpath, and the superintendent's house occupied the highest point overlooking the entire operation. An additional row of upscale houses provided residences for the switching engineer, the mine inspector, and other supervisors. Similarly, Number 2 Town consisted of forty houses constructed in four rows along two streets. Situated between the company store to the north and the entrance to Number 2 Mine to the south, this community represented a quintessential model of late-nineteenth-century company towns.[5]

In fact, mining operations in the Cahaba field generally followed the same developmental pattern manifested in other coal regions. Typically, three types of mines—drift, shaft, and slope—characterized underground mining efforts. Initially, miners dug into outcroppings and followed coal seams along a relatively level path. These drift mines provided easy accessibility, but produced relatively small amounts of coal. In contrast, deeper and thicker seams required more deliberate means of access. Miners dug vertical shafts and used elevators to tap the deepest seams, but slopes offered access to seams closer to the surface. Laying track from the mine entrance to the coal face, workers transported cars along the length of the slope. All three types existed in the Cahaba coal field, but slope mining represented the most common configuration.[6]

Four other Blocton mines—three slope and one shaft—commenced operations by the summer of 1888, and two additional mines were opened during the following year. Number 3 Town developed into a comparatively large community in an abandoned field near Smith Hill. Due to the expansive reaches of the mine itself, miners lived in Number 3 Town, Little Italy, Stack Hill, Smith Hill, and Thomas Town. The mine entrance, located in a draw between two steep hills about one-half mile east of town, provided a challenge for locomotive operators. The switch engine needed a "running start" to push its rail cars uphill to the tipple, and a heavy load or slippery tracks created a ready-made adventure for everyone involved. Other mines developed unique characteristics as well. For example, Number 4 Mine closed by 1890, but the small houses of Number 4 Town continued to provide shelter for several black families. Number 5 Mine comprised the only shaft operation in the Blocton area, and its cage enclosure, boiler, engine rooms, and tipple shed created a distinct silhouette and an unusual atmosphere. Number 7 Mine, a drift mine, used mules to pull loads of coal along the generally level mining floor, and locomotives then transported the coal cars to the Number 2 tipple for weighing and dumping.[7]

As Aldrich and his associates expanded their mining operations, they also diversified their production capability. Building four rows or batteries of beehive ovens southwest of Number 5 Mine, the CCMC commenced coke production. Putting 140 ovens into operation in March 1889, the company doubled that figure by the following spring. By the end of 1890, each battery extended nearly one-quarter mile, and 467 beehive ovens produced six hundred tons of coke daily. The CCMC found a ready market for this fuel, establishing commercial relationships with the Eureka Company at Oxmoor, the Birmingham Furnace and Manufacturing Company at Trussville, and the Pioneer Mining and Manufacturing Company at Thomas near Pratt City. Once absorbed by TCI, the Blocton ovens supplied coke to that company's iron furnaces at Oxmoor and Bessemer.[8]

As coal and coke production increased, Blocton developed into a thriving community. Centrally located, the company store provided a common gathering place for residents of the separate mines and towns. Family dwellings, boarding houses, and commercial buildings lined the streets and gave Blocton a bustling air. Churches, schools, and fraternal organizations fostered a community spirit and helped establish a stable social order. The circus, boxing matches, and civic functions provided entertainment, and baseball teams, community bands, and bicycle clubs offered recreational opportunities.

On the other hand, company towns served multiple purposes. For example, settlements in Pennsylvania's Panther Valley demonstrated the power of place and history. According to Thomas Dublin and Walter Licht, this "historical memory serve[d] a protective function for anthracite region residents." Similarly, Thomas Andrews describes mining towns in Colorado as "political landscapes." Comparable to the settlements constructed in the Carolina Piedmont's textile region, these company-owned and operator-controlled spaces and structures were designed to counter any opposition generated by the pressures and hazards of underground work. In effect, the system preserved unity and order, and company towns embodied the inherent struggle between miners and operators.[9]

Truman Aldrich's brother, William, developed a mining community of his own near the Montevallo mines. William Farrington Aldrich came to Alabama from New York in 1874 to join his brother in forming the Montevallo Coal and Transportation Company. The brothers worked as partners until Truman sold his interest to William in 1880. At that point, William Aldrich transformed the mining camp into a viable community. Constructing a company store near the railroad, Aldrich combined the post office, telegraph office, depot, and meeting hall within one building. In addition, new houses dotted the landscape; rather than constructing typical row housing, Aldrich insisted that "green spaces" remain between dwellings. Furthermore, in an attempt to counter the effects of ever-present coal dust, Aldrich ensured that every house was repainted annually. Varying colors—gray, red, green, and white—gave the housing a homelike appearance. Aldrich also provided for the construction of churches and schools—duplicated to satisfy the needs of the segregated community. In short order, this hands-on operator placed his imprint on the mining community; "Aldrich" became the official name in 1883.[10]

William F. Aldrich began work on a palatial home in 1890, one year after his marriage to Josephine Cables of Rochester, New York. Centrally located, but out of sight of the mines, "Rajah Lodge" consisted of four stories and numerous verandas and balconies. In addition, formal gardens surrounded the home. Designed by a German landscape architect, the grounds incorporated flowers, shrubbery, and fish ponds to form one of the "most magnificent gardens of the South." Four greenhouses contained exotic plants and out-of-season flowers, and a park adjacent to the depot included a rose garden and fountain. In addition to area beautification, Aldrich influenced community spirit and quality of life. The town, which Aldrich claimed was governed primarily by the Golden Rule, became a model of order, peacefulness, and cleanliness. Local historian Henry A. Emfinger credits William Aldrich with typical nineteenth-century entrepreneurial paternalism: "The company would not employ men that were disorderly, nor would they allow whiskey or beer [to be] sold on their property. . . . Aldrich would dismiss any person with questionable character." Hence, the Aldrich community benefited from the presence and influence of its resident owner, but demanding work and elements of social control often generated elements of discontent as well.[11]

These towns, originating in the final two decades of the nineteenth century, paved the way for numerous other communities that dotted the Cahaba landscape after 1900. Moreover, Blocton and Aldrich typify the settlements described by Crandall A. Shifflett in *Coal Towns: Life, Work, and Culture in Company Towns of Southern Appalachia, 1880–1960*. This definitive work traces the growth of more than five hundred coal towns that began in the 1880s and peaked in the 1920s. Shifflett identifies three distinct periods of development. "Pioneer Days" defines

Fig. 11. "Rajah Lodge," home of William F. and Josephine C. Aldrich. (Courtesy of Aldrich Coal Mine Museum.)

the early mining period from the 1880s to 1915. The "Paternalistic Period" spans World War I and the Great Depression (1915–1930). And "Decline" marks the period from 1930 to the 1960s.[12]

Shifflett's pioneer period places coal-mining operations within a rugged frontier venue. In the late nineteenth century, poor working conditions and low quality of life stemmed from an austere, Spartan-like environment. A male workforce formed makeshift mining camps consisting of temporary housing such as tents and tar-paper shacks. However, as the early days of Blocton and Aldrich illustrate, Shifflett's model does not apply universally. Coal operations began at Piper in 1901 and at Margaret in 1905, and numerous other communities developed during the next three decades. Still, Shifflett's periodization and overall description of Appalachian coal towns proves instructive when surveying Cahaba. Former residents of Piper contend that "Piper was different from other mining towns because we loved one another and shared what we had during hard times." Similarly, inhabitants of Margaret describe their town as "a unique, happy community." In fact, every town had its share of unique characteristics, and Shifflett is correct when he asserts that "no single town represents all coal towns." Thus, an examination of select settlements provides an important corrective to generalizations about life among the coal miners of the Cahaba field.[13]

Coal production in the Cahaba field paralleled that of other Appalachian regions. Overall output tripled in the final decade of the nineteenth century and increased fivefold by 1930. This meteoric rise in production stemmed from the nationwide transition from wood to coal as the primary source of energy. Significant increases in demand began in 1885, and the United States marked century's end as the world's leading coal producer. In addition, bituminous coal and coke replaced anthracite as the fuel of choice for the nation's industries. By 1930, Appalachian mines produced 80 percent of the bituminous coal extracted in the United States.[14]

Increased demand for coal coincided with an agrarian crisis that prompted workers to migrate from farming areas to relatively remote mining sites. Shifflett suggests that this radical shift in occupations held no long-term negative implications for these displaced families. Rather, he contends that family units brought their rural folkways with them and that these values and traditions helped to define a sense of community based on kinship. This communal setting fostered a sense of culture and social life linked by collective identity, mutual assistance, and solidarity. Consequently, Shifflett reports that "many former mining-town residents . . . recalled with fondness their days in the company town setting, in spite of the noise, poor sanitation, dust, danger, and company authority." In effect, life on the farm offered little promise, so families often moved to the coal towns in search of economic opportunity and a better way of life. Alabama historian Wayne Flynt supports this argument when he posits that, for most families, "life in a coal camp was better than what they left behind."[15]

In general, geological folds and faults divide the Cahaba field into three distinct coal basins. The Henry Ellen Basin lies to the north, the Blocton Basin (including the smaller Acton and Cahaba Basins) defines the southwestern portion, and the Montevallo Basin encompasses several seams to the southeast. Multiple veins of coal run through one or more basins, and these seams determine the best locations for mining operations. The Montevallo Seam (approximately 3.5 feet thick) lies closest to the surface. The Helena (3.75 to 7 feet) and Thompson/Underwood (4 to 5 feet) Seams underlie the Montevallo. Next, the Henry Ellen (or Mammoth, Black Shale, Gholson) Seam measures 11 feet. The Buck (or Clark, Woodstock, Blocton Number 1; 3 feet), the Youngblood (2.75 to 4.5 feet), the Nunnally (2.5 to 5 feet), and the Harkness (4.75 feet) lie beneath the Henry Ellen. Layers of shale and sandstone separate the seams.[16]

During the hand-loading period, miners entered the mines equipped with the tools of their trade—picks, shovels, pry bars, breast augers, saws, axes, and tamp-

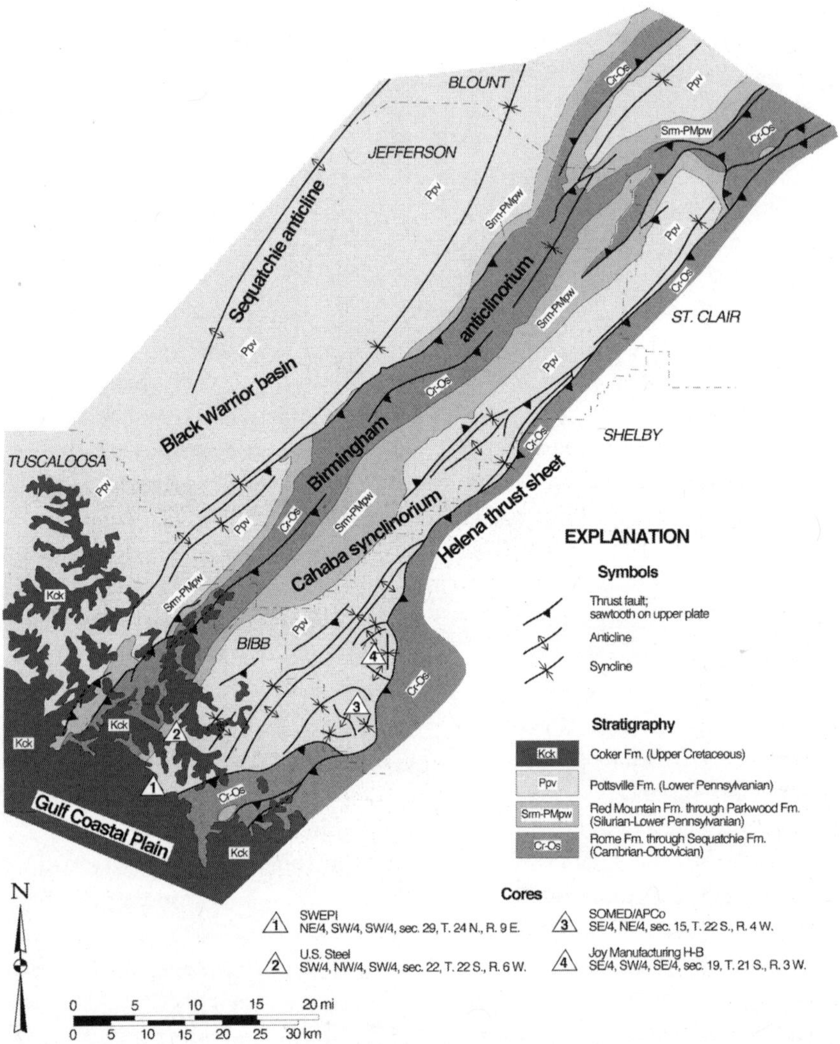

Fig. 12. Generalized geologic map of the Cahaba synclinorium and adjacent areas. The Cahaba coal field extends along the southeastern margin of the Cahaba synclinorium and is defined by the outcrop of the Lower Pennsylvanian Pottsville Formation. Consisting of three major basins, several coal seams, and numerous folds and faults, the Cahaba field follows the Cahaba River Valley through portions of Bibb, Jefferson, Shelby, and St. Clair counties. (Jack C. Pashin, Richard E. Carroll, Robert L. Barnett, and Mirza A. Beg, *Geology and Coal Resources of the Cahaba Coal Field*, Bulletin 163 [Tuscaloosa, AL: Geological Survey of Alabama, 1995], 3.)

ing bars. In addition, they carried three-tiered dinner buckets that contained their food and drinking water. Water filled the larger bottom section, food occupied the middle tray, and dessert rested in the top portion. Also, kerosene oil lamps provided dim, smoky lighting. Small metal containers (approximately three or four inches high) contained the fuel. A spout with a wick and a hook (used to attach the lamp to a soft cloth cap) protruded from the container. After the turn of the twentieth century, carbide lamps replaced the kerosene lights. The bottom chamber contained small pieces of carbide, and water filled the upper chamber. A metal arm regulated the flow of liquid, and the combination of water and carbide produced a gas. Miners rotated a flint wheel (similar to a cigarette lighter) to ignite the gas as it emitted through a reflector. Later, in the 1930s and 1940s, battery-powered lamps eliminated the need for an open flame.[17]

Once at the coal face, miners began the four-step mining process: undercutting, drilling, blasting, and loading. First, they undercut the coal seam by carving a three- to four-foot wedge into the lower part of the face. Using picks, miners usually performed this function on their knees or lying on their sides. With the undercutting completed, miners drilled holes in the coal face using a 5.5-foot-long auger with a U-shaped crank. Pressing their breastplates against the augers, miners determined the location and angle of each hole to maximize the effect of the blast. Inserting a thin roll of waxed paper containing black powder, miners tamped the charge with dampened dirt. Then, after penetrating the charge with an iron rod or needle, they inserted a "squib"—a thin roll of waxed paper with powder at one end that served as a fuse. Yelling "fire in the hole" to alert other miners as they lighted the fuse, miners ran for cover to await the blast. After "shooting" the coal, miners returned to the face to load the coal into trams or cars, each with a one- to two-ton capacity.[18]

Separating the shale and rock, miners used shovels to load the coal. Marking each filled car with a brass tag, miners formed a five- to ten-car "coal trip." Pulled by mules and hoisted by wire ropes or cables, the cars traversed the tracks to the mine entrance. At the tipple, hickory "sprags" immobilized the wheels, checkweighmen weighed each car and credited tonnage to respective miners, and "day men" dumped the coal. Mechanical shakers separated large lump from medium lump coal; nut coal and steam or slack coal were also separated.[19]

Miners received payment solely for coal tonnage. Consequently, other necessary functions such as laying track, timbering roofs, maintaining tools and lamps, rolling powder, and mending clothes constituted "dead work." Later, as miners organized for collective action, payment from "portal to portal" became a critical issue. Once approved, miners would be paid for transportation time between the mine entrance and the coal face as well as for "dead work." "Company men" or "day men" performed support tasks inside and outside the mine and received an hourly or daily wage. Slate picking and various top-side duties such as carpentry, masonry, electrical work, and engineering constituted support functions.[20]

The "room and pillar" method represented the standard process for extracting coal during the hand-loading period. Adopted in the nineteenth century, this technique persisted until the introduction of mechanical loaders in the 1930s. Operators opened two parallel openings or "butt headings," each ten to twelve feet wide. Situated approximately thirty feet apart, one contained track for transporting coal cars, and the other provided ventilation. Rooms opened at right angles from the main headings, and miners tunneled into the coal seam at a rate of ten to twenty feet per day. Solid walls of coal, or pillars, separated each room, but "breakthroughs" or "cross-cuts" enhanced ventilation and fostered communication among adjacent miners. Rooms averaged twenty-four feet in width and were as high as the seam of coal (usually two to eight feet). Sixty feet, measured from center to center, separated adjacent rooms. When rooms were "mined out," miners attempted to remove the remaining coal by "pillar robbing." This technique proved dangerous inasmuch as miners allowed the unsupported roof to collapse onto the mine floor.[21]

Until mechanization transformed the coal industry in the 1930s and 1940s, the room and pillar method constituted the most efficient mining technique. According to industrial archeologist Jack Bergstresser, "no other system of mining would produce as clean a coal as this meticulous hand mining process." Even machine mining, although faster, produced "dirty" coal and required washing after extraction. The washing process took advantage of the scientific principle of specific gravity. After being segregated by size, either by screening or crushing, "run-of-mine" coal was placed in a washing jig. The specific gravity of the coal and the accompanying impurities (e.g., bone, slate, shale, pyrite) caused these products to separate when immersed in water. In this case, mechanical washers significantly reduced the ash content and made Cahaba coal suitable for coking blends. Washed coal sold for 15 to 50 cents more per ton than run-of-mine or dirty coal. In addition, the cleaner coal reduced handling and hauling costs by removing the impurities early in the process. Overall, a 5 percent reduction in the ash content would reduce the ultimate cost of pig iron production by $1 per ton. Thus, by washing coal, operators sought "maximum reduction of impurities and maximum recovery of good coal with a minimum loss of good coal with the refuse."[22]

By 1925, nearly two-thirds of Alabama miners lived in company towns. According to a federal study, "life in a coal-mining community . . . follows a routine. . . . The doings of the miner, his wife, and his children are synchronized with the action of the tipple." Moreover, Shifflett asserts that "coal towns in southern Appalachia were largely the creations of the companies that secured the leases, founded the towns, and hired the miners, i.e., they were 'company towns.'" Operators owned

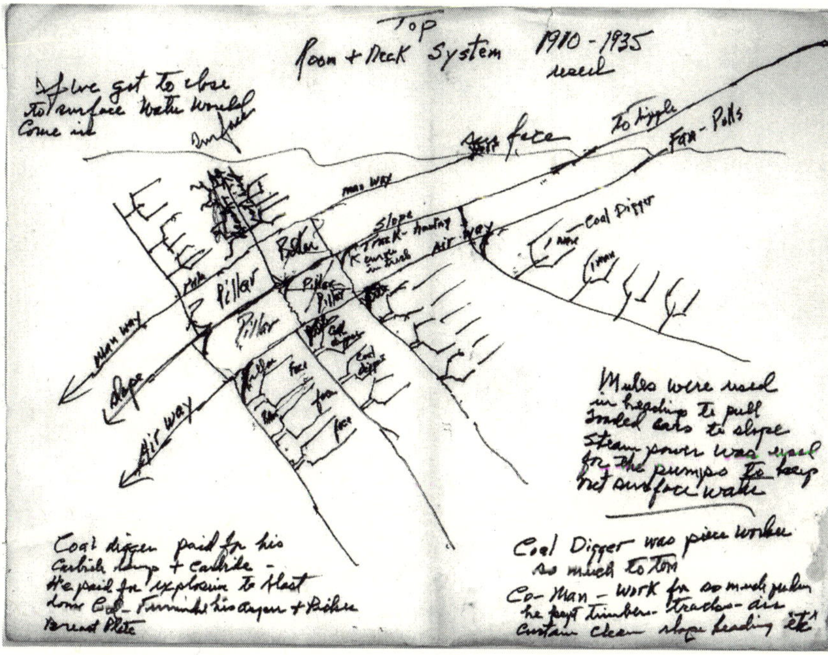

Fig. 13. Mining sketch of the room and pillar method of extraction. (Drawing by Jess Shepard of Newnan, GA. Courtesy of the author.)

the land, shops, stables, drum houses, tipples, and all of the infrastructure. In addition, mining companies provided houses, schools, churches, commissaries, theaters, dancehalls, bowling alleys, soda fountains, graveyards, and other necessities for their workers. In sum, "the company owned everything."[23]

Ordinarily, responsibility for naming each coal town rested with the operator. As exemplified by Blocton and Aldrich, many towns were named by or for their owners. Some towns derived their names from a respected engineer, general manager, or company official, whereas others reflected the names of revered women. As I. T. Tichenor predicted in the late 1880s, the Cahaba field spawned several mining communities whose names bear testimony to their origins.[24]

Various rail lines crisscrossed the 270 square miles of surface area, connecting mines to markets. Blocton, centrally located among Bibb County mines, was serviced by a half dozen railways—the Alabama Great Southern, the Mobile & Ohio, the Louisville & Nashville (L&N), the Southern, the Woodstock & Blocton, and the Birmingham Mineral. Ultimately, hundreds of Blocton miners opened a dozen slopes to tap the Woodstock and Thompson Seams. Using mules to haul coal underground, operators used steam- or electric-powered hoist mechanisms to carry the coal to the surface. Large ventilation fans, driven by steam or electricity, kept fresh air flowing through the slopes and splits. Operating for 236 to 281 days per year, miners earned 57.5 cents per ton in 1913 and received their highest rate of $1.09 per ton in 1920. The Hill Creek Coal Company, TCI, the Alabama-Blocton Coal Company, and the Black Diamond Coal Mining Company produced lump, slack, and run-of-mine coal until the Number 9 Mine closed in 1958.[25]

The Southern Railway connected three Bibb County mines: Hargrove, Klondyke (named by two miners who participated in the Canadian gold rush of 1897–1898), and Leland (named for George Randall's second son). At Hargrove, miners for the Cahaba Southern Coal Mining Company extracted an annual average of sixty thousand tons of lump and nut coal from the Thompson Seam from 1913 to 1918. Using mules, a gasoline-powered locomotive, an electric hoist, and a steam-driven fan, 34 to 142 miners worked 272 days per year. A similar configuration at Klondyke (or Smith's Shaft) produced 20,005 tons of lump, slack, and run-of-mine coal in 1929 and an annual best of 67,378 tons in 1936. Tapping the Woodstock Seam, 48 to 168 miners of the Blocton Mining Company and Howell's Mining Company worked 183 days per year and earned between 80 cents and $1 per ton. At Leland, the Cahaba Central Coal Company mined the Younglood Seam for 160 days, producing fourteen thousand tons of run-of-mine coal in 1914.[26]

The Mobile & Ohio Railroad connected Kyomie, Norris, Red Eagle, Lucile, Braehead, and Dixie. When operations closed at Kyomie in 1913, miners employed by the Kyomie Coal Company had produced six thousand tons of run-of-mine coal from the Helena Seam. At Norris, the Bibb Coal Company mined the Yessick Seam from 1918 to 1923. Employing thirty-eight to seventy-five workers and paying them $1.14 per ton in 1920, Bibb Coal produced 30,074 tons of run-of-mine coal in 285 days. The Red Eagle (or Weatherford) Coal Company extracted all types and sizes of coal from the Woodstock Seam until 1922. Miners earned from 57.75 cents to $1.09 per ton, working an average of 252 days per year. Red Eagle employed its most workers (137) in 1916 and realized its largest output (91,735 tons) two years later. Lucile, named for Henry DeBardeleben's sister, used convicts and free laborers to produce lump and run-of-mine coal until 1929. Operating for 230 days per year, the Red Feather Coal Company opened two slopes and six drifts along the Thompson Seam and used a Montgomery washer (four-hundred-ton capacity) to generate a cleaner product. Paying miners 57.75 cents (1913) to $1.09 (1920) per ton, this operation employed an average of 164 workers. Braehead, derived as a composite using the Scottish "brae" for hill, suggests a dominant Scottish or northern English population. However, this mine operated for only two years (1922–1923). At Dixie, the Moffat Coal Company commenced operations in 1922 and began using an electric chain machine the following year. Employing a maximum of 450 workers in 1927, this mine operated for 273 days per year. Initially producing run-of-mine coal, Dixie converted to lump, nut, and slack production in 1926. Relying on mules and an electric hoist to bring the coal to the surface, operators used an electric motor to transfer loads to the tipple located three thousand feet from the slope.

The Birmingham Mineral Railroad ran from Gurnee Junction to Blocton and linked mines at Savage Creek, Youngblood, Marvel, Garnsey, Coleanor, Piper, Belle Ellen, and Lucile. Cyrus Garnsey, a capitalist from Memphis, organized the Choctaw Coal & Mining Company in 1913 and opened mines at Garnsey and Savage Creek. Subsequently owned by the Galloway Coal Company of Memphis (1913–1920), the Garnsey Coal Company of Birmingham (1921–1924), and the McSherry Coal Company of Marvel (1928), the Garnsey Mine employed 140 to 339 workers and produced 22,236 to 37,611 tons of lump, slack, and run-of-mine coal each year. This operation on the Thompson Seam used mules, a gasoline-powered locomotive, a steam-powered fan, and a seven-hundred-ton Montgomery washer. From 1915 to 1918, the Bessemer Coal, Iron, & Land Company tapped the Youngblood Seam, using a configuration similar to that at Garnsey. Employing an average of ninety-three workers and paying miners 57.75 cents to $1.00 per ton, the Youngblood Mine extracted all types and sizes of coal. Working for 264 days in 1917 and 265 days in 1918, miners produced 76,540 and 70,027 tons, respectively.

At Belle Ellen (named for Henry DeBardeleben's daughter, Belle, and his wife, Ellen) the Bessemer Coal, Iron, & Land Company opened six mines on the Youngblood, Woodstock, and Thompson Seams. Convicts mined coal at Number 2 Mine and Number 5 Mine, and free miners earned 57.75 cents to $1.09 per ton. Belle Ellen enjoyed its "Golden Age" during the 1920s, producing lump, slack, and run-of-mine coal; Number 5 Mine continued operations until 1950.

The Marvel Mines and community represent the efforts of B. F. Roden and the Roden Coal Company. Historian Henry M. McKiven Jr. mentions Roden in relation to the Avondale Land Company in his study of the iron and steel industry of Birmingham. Roden sold lots to local workers and merchants and also invested in a rail line from Birmingham to Avondale to facilitate commuting. His mines, based on a British model and configured with one entrance (Number 1 Mine) directly above the other (Number 2 Mine), gained the reputation of being state-of-the-art. These mines operated on identical plans, opening similar entries and rooms and sharing tipple, washer, and ventilation systems. Tapping the Clark and Gholson Seams and converting to machine mining in 1934, these mines produced lump, nut, and slack coal until the Number 1 Mine ceased operation in 1954. A Number 3 Mine opened in 1917, but 1913 marked the company's greatest production—283 days worked and 347,633 tons of coal produced.[27]

Henry F. DeBardeleben initiated another independent mining operation when he convinced Memphis businessman Oliver Hazzard Perry Piper to invest in the

Fig. 14. Roden Coal Company, Marvel, Alabama. (Courtesy of Marshall Goggins.)

Fig. 15. Marvel Mines Nos. 1 and 2. "State-of-the-art" mines with one entrance directly above the other. (Courtesy of Marshall Goggins.)

Little Cahaba Coal Company (LCCC) upon its inception in 1901. Born in Maryland in 1838, Piper moved to Memphis with his two brothers in 1859. Having served in the Confederate Army during the Civil War, he made money by shipping sand along the Mississippi River; his other business ventures included brick making, contracting, and building. Investments in some Louisiana sulfur mines increased his wealth and led to his involvement in financial and commercial affairs. Major undertakings included construction of the Nashville, Chattanooga, & St. Louis Railroad as well as the Frisco Railroad between Memphis and Birmingham. At the turn of the twentieth century, Piper served as president of the Bank of Commerce and Trust Company in Memphis.[28]

While visiting in the home of John Charles Henley, vice president of the First National Bank of Birmingham, Piper asked Henley's son, Walter, to oversee his interests in the mines. Henley agreed and became secretary-treasurer of the company. Thus, Piper became an absentee owner, and the other major stockholder, Joseph Riley Smith Jr., served as president. Smith's father, Dr. Joseph Riley Smith Sr., was a stockholder and director in the First National Bank of Birmingham and one of the wealthiest men in Alabama. J. R. Smith Jr., who inherited part of his father's fortune and owned several ore mines outright, remained as president for three years before relinquishing that office to Walter Henley. Henley served as company president until his election as president of Birmingham Trust National Bank in 1925.[29]

During his tenure, Henley directed mining operations along the Thompson Seam near the Civil War–era Thompson mines. After incorporating the LCCC in January 1901 with $50,000 in capital stock, DeBardeleben and Henley initiated the extension of the Southern Railroad line from Seymour. Having thereby established contact with Birmingham, the LCCC shipped its first load of coal in October 1901. Three years later, the LCCC expanded its operations by opening a second slope mine. Extending the railroad line from Piper Number 1 to Piper Number 2, the LCCC increased its production to four hundred tons per day. With a seam thickness exceeding twenty-five feet, Piper Number 2 became the most productive mine under LCCC control. Massive amounts of coal often forced miners to build scaffolds along the face to place their powder charges at the appropriate levels. Advertising their lump coal as "the South's Best" for steam or heating, workers in Piper Number 2 frequently claimed that "Piper coal was of such good quality that you could almost set it on fire with a match."[30]

In 1905, the LCCC purchased the Coleanor (a combination of coal and Eleanor) Mine from the Blocton Cahaba Coal Company. This enterprise, begun in 1901 by Jesse B. Wadsworth and Fred M. Jackson, operated a single slope mine. Having purchased a seven-hundred-acre area, Wadsworth and Jackson completed rail spurs from the Southern and L&N Railroad lines and commenced operations

Fig. 16. Coleanor Mine, owned by the Blocton Cahaba Coal Company and Little Cahaba Coal Company. (Courtesy of Marshall Goggins.)

in July 1901. Constructing an additional line between Coleanor and Piper, the expanded LCCC managed three mines—Coleanor, Piper Number 1, and Piper Number 2. Converting to a steam-powered electric generator after acquiring the Coleanor Mine, the LCCC supplied electricity to its mining operations and to the miners' homes. The company also converted to machine mining in 1925, producing lump, nut, and slack coal after the mid-1930s. With approximately five hundred workers employed by the company, the total population of Piper and Coleanor peaked between 2,500 and 3,000. These combined communities consisted of 565 houses, three churches, a school, three boarding houses, a barber shop, a dentist office, and two doctor's offices.[31]

The abundance of mines in Bibb County illustrates the overall nature of operations within the Cahaba field. Instead of experiencing centralized control and standardized wages from one or two large corporations, Cahaba miners dealt with numerous independent operators. Consequently, the number of days worked, pay rates per ton, and overall output varied among the smaller enterprises of Cahaba.

In Shelby County, three rail lines—the Atlanta, Birmingham & Atlantic, the L&N, and the Southern—connected coal towns. The Atlanta, Birmingham & Atlantic linked the old Eureka Mines and the Paramount Mines with the

L&N Railroad near Helena. Both operations dealt with coal seams that pitched and rolled at a twenty-two- to twenty-three-degree incline. The Eureka Company purchased the Rolling Mill Mine from the Connors-Weyman Steel Company of Birmingham in 1916. Opening four slopes on the Helena and Thompson Seams, the Eureka Company's operation used mules, electric hoists and fans, and a five-hundred-ton American Concentrated Company washer to produce all types and sizes of coal. The Number 2 Mine averaged seventy-one workers for 225 days per year and produced 124,654 tons in 1916. The Number 3 Mine employed 124 workers for 183 days per year and extracted 162,285 tons in 1918. At the Number 4 Slope, seventy-eight workers mined coal for 190 days per year and produced 60,922 tons in 1920. Miners earned 52.5 cents to $1.04 per ton, but operations ceased in 1927. The Paramount Coal Company operated four slopes from 1922 to 1948. The Number 1 and Number 3 Mines tapped the Helena Seam, Number 2 Mine extracted from the Thompson Seam, and Number 12 Mine used long-wall machine mining on the Coke or Black Shale Seam. Numbers 1 and 2 shared a tipple and used a Fordson tractor to transport coal from the slopes. The Number 12 Mine trucked its coal to a preparation plant approximately two miles away. An annual average of fifty-three miners earned 70 cents to $1.00 per ton, worked 199 days, and produced 31,178 tons. The Paramount operation mined run-of-mine coal into the 1930s, but produced lump, nut, and slack coal after World War II.[32]

The Southern Railway linked several mining communities in the southeastern portion of the Cahaba field. The Southern Coal & Coke Company began operations at Gurnee Junction, working the Glen Carbon Mine on the Gholson Seam until 1916. Miners earned 57.5 cents per ton, worked 204 days per year, and produced an annual average of 44,947 tons of run-of-mine coal. Expanding operations to include the Clark Seam, the Southern Coal & Coke Company eventually opened four slopes at Boothton. These coal mines employed an average of 210 workers per year until 1929, producing 189,970 tons of all types and sizes in that year. Employment decreased to 105 workers in the 1930s and 1940s, and the Boothton mines averaged 202,304 tons per year in the 1930s. After World War II, the annual average increased to 315,537 tons, but coal production ended in 1951.

The Southern Coal & Coke Company operated the Superior Mines until 1913, and the Shelby Coal Company at Sicard reopened the mines in 1921. Working on the Helena Red Ash Seam, nineteen workers averaged 147 days per year and produced 9,704 tons of run-of-mine coal annually. Superior closed in 1923, and the Peerless Cahaba Coal Company at Straven opened the nearby Peerless Mine two years later. Operating for five years on the Helena Seam, day laborers, company workers, and contract miners worked 198 days per year and produced 77,185 tons of lump, nut, and slack coal. The Peerless Company used long-wall machines, mules, an electric hoist and fan, and a Montgomery washer. Similar operations by the Climax Coal Company at Maylene extracted lump coal for domestic use. An average of seventy-two miners worked 237 days per year and earned 90 cents to $1.55 per ton. Production of 29,663 tons made 1917 the best year for the Climax Company.

Two mining communities—Sicard and Straven—provide a classic example of the opportunities available for the relatively small investor. Adrien Sicard, the younger of two sons, was born in 1869 in the village of Pouzol in southwestern France.[33] Due to the custom of primogeniture, his brother inherited the family land, homestead, and vineyard upon their father's death during the Franco-Prussian War (1870–1871). To avoid compulsory military service, Adrien and his cousin, Paul Faurie (both age seventeen), left France for the United States. Arriving in New York City, Sicard headed west to the Illinois wheat fields, and Faurie acquired a construction job in New Jersey. Ill-suited to the life of midwestern farmers, Sicard left Illinois for Pittsburgh where he gained employment as a coal miner. Working his way to foreman, he learned many of the details of the industry. Having heard about Birmingham, the "Pittsburgh of the South," Sicard headed to Alabama. As foreman of a mining operation at Blossburg, he developed many innovative techniques and earned a promotion to superintendent. After a visit to France in 1900 and his marriage a year later, Sicard operated a captive mine at Helena for the Conners-Weyman Steel Company of Atlanta, Georgia. Meanwhile, he opened a mine in North Birmingham under the name of the Federal Coal Company.

Striking out on his own in 1917, he opened a new mining camp in the vicinity of Helena, named Sicard. Beginning with a sawmill, Sicard constructed two-, three-, and four-bedroom houses, a commissary, two combined school-church buildings, a doctor's office, and an engine room. The wet, level coal seam allowed the use of carbide lamps as there was no threat of gas buildup or explosion. Sicard paid royalties to TCI, owner of the mineral rights, and sold his coal to the Southern Railroad. His oldest daughter, Mary, operated the commissary and also served as company bookkeeper.

With Sicard in full operation, he opted to purchase the nearby Straven Mine from the Peerless Cahaba Coal Company. One of the largest mining operations in Alabama at that time, Straven paid royalties to the Southern Mineral Land Company. Sicard moved in and immediately added enough houses to accommodate approximately a thousand inhabitants. This area, known as New Straven, included an office building, drugstore, recreation center, pool hall, doctor's office, and homes for the company executives. Unlike Sicard, the Straven Mine did contain gas, and an explosion occurred on January 13, 1926. Unable to use dynamite, black powder, or other explosives, Sicard traveled to the German-occupied regions of Alsace and Lorraine to study coal-cutting machines and to learn mechanized mining techniques. Even though he introduced some mechanization into his Shelby County operations, the Sicard mine closed in 1927, and Sicard's wife sold the Straven Mine after his death in 1932.[34]

Tapping the Helena and Montevallo Seams, Adrien Sicard's mining operations produced all types and sizes of coal. At Sicard, miners earned 68 cents to $1.09 per ton, and fifty-nine workers operated 199 days per year to produce an annual average of 30,520 tons of run-of-mine coal. At Straven, eighty-five day laborers and free miners produced 37,283 tons of lump, nut, and slack coal in 225 working days. In the best year, 1924, 181 employees worked 264 days to extract 87,075 tons.

Farther to the south, mining operations at Dogwood and Aldrich produced lump, nut, and slack coal well into the 1940s. Both operations converted to machine mining along the Montevallo Seam and used gasoline- and electric-powered locomotives as well as electric hoists and fans. At Dogwood, the Montevallo Domestic Coal Company (1915–1917) and the Little Gem Coal Company (1918–1948) used a Fordson tractor to transport coal to the tipple and a two-hundred-ton Montgomery washer to clean the run-of-mine coal. Dealing with a twelve-degree pitch, Dogwood miners earned 50 cents to $1.50 per ton and worked an average of 240 days per year. Average annual output equaled 13,411 tons before mechanization and 87,664 tons thereafter. At Aldrich, a workforce averaging 298 convicts (1915–1927) or free miners worked 302 days per year, and production reached new highs in 1917 (157,372 tons), 1926 (163,634 tons), and 1927 (178,564 tons).

The L&N Railroad ran through Mossboro, Coalmont, and Acton. At Mossboro, the Premier Coal Company employed fifty-nine workers to mine the Black Shale Seam from 1922 to 1934. Miners earned 65 to 90 cents per ton, worked an average of 204 days per year, and produced 31,752 tons annually. The Helena Mining Company operated the Ebony Mine along the Thompson Seam at Tacoa until 1917. The Brookside-Pratt Mining Company bought the mines at that time and continued operations until 1922. Known as the Blue Diamond Mines, four slopes tapped the Clark and Gholson Seams (Coalmont Number 1), the Clark Seam (Number 2), the Helena Seam (Number 3), and the Thompson Seam (Number 4). Using mules, electric hoists and fans, and a six-hundred-ton Montgomery washer, 108 workers produced 69,273 tons of coal per year. Miners earned 52.5 cents to $1.09 per ton as they extracted lump and nut coal from the mines. Moreover, Number 4 Mine used chain machines to undercut the coal and used electric lighting to illuminate the slope.

The L&N extended spur lines to Acton in 1906 when Henry F. DeBardeleben opened Acton Mines Number 1 and Number 2. After his ouster from TCI in 1894, DeBardeleben organized the Alabama Fuel and Steel Company, initiated coal-mining operations in southwestern St. Clair County, and opened two mines at Margaret in 1905. In 1908, the company's name changed to the Alabama Fuel and

Fig. 17. Advertisement for the Montevallo Coal Mining Company, Aldrich, Alabama. (Courtesy of Henry Emfinger and Aldrich Coal Mine Museum.)

Fig. 18. Charles F. "Uncle Charlie" DeBardeleben, coal-mining enthusiast and entrepreneur. (Marie Butler, *Margaret, Alabama . . . and Now There's Gold: A Brief History of a Unique Community* [Birmingham: privately printed, 1989], xi.)

Iron Company (AF&IC) with J. M. Overton as president, H. F. DeBardeleben as vice president, and C. F. DeBardeleben as vice president and general manager. At this juncture, DeBardeleben added mines, a power plant, and a coal washer at Acmar. Furthermore, AF&IC operations established a link among coal-mining operations in Shelby, St. Clair, and Jefferson Counties.[35]

Upon his father's death in 1910, Charles F. DeBardeleben inherited leadership of the AF&IC. Promoting "true Americanism" and adopting the social welfare capitalism that flourished during these years, "Uncle Charlie" became known as everyone's friend. Margaret historian Marie Butler reports that DeBardeleben made the welfare of his employees and their families his top priority and that he fostered "a harmonious relationship between the employer and employee." However, historian Ronald L. Lewis contends that the DeBardeleben brand of paternalism constituted a two-edged sword. Three generations of DeBardelebens—Henry, Charles, and Prince—exercised paternalism with a purpose. By providing basic services for their employees, these magnates determined to build a fiercely loyal, antiunion workforce. To ensure compliance with company policies, the DeBardelebens established elaborate spying networks and immediately fired anyone who joined the union. During World War I, Charles DeBardeleben personally canvassed miners at Margaret and encouraged those who habitually worked two days per week to increase their availability threefold. According to "Uncle Charlie," miners should work, "not for the money . . . but because the government needs the coal." Thus each miner's productivity became the measure of devotion to his patriotic duty.[36]

The AF&IC opened two mines at Overton (Jefferson County) in 1921, but President J. M. Overton died in an automobile accident shortly thereafter. DeBardeleben assumed total control of the company's mines at Margaret, Acton, Acmar, and Overton. Each of these towns represented a self-sufficient community with mines, tipple, houses, commissary, industrial shops, warehouses, and schools, but a company-owned railroad provided corporate and social links. DeBardeleben continued to expand the company's operations so that by 1935 the AF&IC constituted the state's largest producer of commercial coal.[37]

Four Acton mines tapped the Helena, Black Shale, and Thompson Seams, producing all types and sizes of coal. An average of eighty-nine workers at Number 1 Mine and sixty-seven workers at Number 2 Mine worked 242 days per year and extracted 114,210 tons in 1914. Miners earned 50 to 99 cents per ton, but production decreased to 80,253 tons in 1925. The AF&IC's Acton operations ended with the closure of Number 2 Mine in 1928, but some independent mining continued along the Helena Seam and within the Acton Basin until 1935.

Branches of the Central of Georgia Railroad extending from Henry Ellen serviced the mining communities of St. Clair County, and the Seaboard Air Line Railway provided passenger service through Parsons and Sanie. At Margaret, AF&IC miners opened seven slopes along the Harkness Seam, but had to deal with gas in Numbers 4, 5, 6, and 8. Inspectors classified each of the four slopes as "gaseous" and "closed light mines." In fact, local regulations stipulated that "all men working inside are required to use the approved type of electric safety lamp and no matches or open lights are permitted." Payment rates varied from 50 cents to $1.015 per ton, but usually ranged between 72 and 84 cents. Average output from 1922 to 1929 equaled 336,349 tons of lump and slack coal, and several slopes shared ventilation fans and tipples. For example, a "dinkey" locomotive transported coal from the Number 4 Mine to the Number 1 tipple, and coal extracted from the Number 6 and Number 8 Mines was hauled 1.5 miles to the Number 3 tipple and an eight-hundred-ton Colgate washer. At Acmar, miners opened eight slopes on the Henry Ellen Seam, extracting lump coal along with some nut, slack, and run-of-mine. Miners worked an average of 260 days per year, earning 54 to 64 cents per ton. Production peaked in 1926 with 463,563 tons, and operators converted to machine mining in 1934. Operations ended in 1951 when Margaret Number 6 and Acmar Number 5 closed.[38]

The AF&IC also opened mines at Henry Ellen and New Acton. Operating at Henry Ellen from 1922 to 1929, thirty-six workers averaged 227 mining days per year. Miners earned 80 cents to $1.00 per ton and produced 17,052 tons of lump, nut, and slack coal per year. Tapping the Helena Seam at Acmar, five New Acton slopes produced lump and nut coal through the 1920s and then added slack coal in the 1930s. Miners earned 55 cents to $1.09 per ton and worked 270 days per year. After converting to machine mining in 1926, New Acton's average annual production equaled 85,538 tons. Output jumped a decade later to 201,085 tons (1936), but the closing of Mines Number 3 and Number 5 two years later terminated operations.

The AF&IC's Jefferson County operations at Overton (near Irondale) tapped the Lower Nunnally (or Bragg) and Upper Nunnally (or Glass) Seams. Operations began in 1908, and Henry DeBardeleben named the coal town for Jesse M. Overton, the Nashville capitalist who financed the venture. The AF&IC built a commissary, and the community developed along both sides of Overton Road. Black and white company housing, schools, and churches provided "separate but equal" facilities in the segregated town. Converting to machine mining in 1924, Overton workers averaged an annual output of 130,235 tons from Number 1 Mine and 179,835 tons from Number 2 Mine. Earning 57.5 to 72.5 cents per ton, miners extracted lump, nut, and some slack coal. Mines Number 3 and Number 4 opened in 1927 and 1934, respectively, but underground operations ceased in the mid-1930s. Some strip mining occurred during the mid-1940s, but all production ended in 1953.

Other mining operations in Jefferson County followed a relatively narrow line of faults and folds along the Cahaba River valley. The Rowen Mine, one of the oldest in the county, opened in 1861–1862 to supply coal to the McElwane Furnace near Gate City. This drift mine followed an unknown seam that outcropped along the riverbank. Jesse Glass briefly operated the mine in 1913, producing 450 tons in 80 days. A. B. Ratliff opened a drift mine along the Ratliff Seam in 1903 and continued sporadic operations until 1929. Hauling coal by wagon for 1.75 miles to Lovick, Ratliff averaged 243 workdays in each year of operation. During the 1920s, he extracted an average of 2,772 tons of run-of-mine coal. Similarly, W. S. L. Pool operated a drift mine along the Margaret Seam from 1908 to 1913. Based seven miles from Irondale along the Cahaba River, Pool hauled wagonloads of coal to Irondale or Gate City. In addition to these wagon mines, the Shades Valley Coal Company opened a slope on the Bragg Seam in 1922. With an average annual output of 41,114 tons, the Shades Valley Mine combined with the Overton operations in 1927.

Finally, the Birmingham Water Works Company conducted mining operations on the Henry Ellen Seam in eastern Jefferson County. With no railroad connection, miners used mules to convey coal to the surface, then a "dinkey" locomotive transported the coal for 1.5 miles to a pumping station. Miners worked

278 days per year and produced an annual average of 20,929 tons of run-of-mine coal. During their best year, 1920, miners extracted 30,980 tons and earned $1.09 per ton.

Meanwhile, Blocton continued to flourish. As mining operations expanded, other settlements developed in the Blocton area. Number 1 Town and Number 2 Town thrived, and Number 3 Town sprang up near the Odd Fellows Hall. Other communities developed on Smith Hill and Stack Hill where houses were built of logs and chinked with clay. Thomas Town emerged when approximately twenty families settled along the road to McCulley Hill and Belle Ellen. Campbelltown consisted of about twenty houses near Number 7 Mine. John T. Milner opened the Hill Creek Mine in 1908, and McGraw Town developed around the commissary, tipple, and coal washer. With a good water supply, gardens and fruit orchards, and proximity to the local schools, McGraw Town became "one of the most desirable places in the Blocton District in which to live."[39]

In addition, black miners settled in Bucktown to the west and in Number 4 Quarter near the abandoned Number 4 Mine. According to Charles Adams, Bucktown's forty houses served as Blocton's "exclusive Negro quarter." European immigrants also settled in Blocton, thereby creating a three-way system of segregation. Immigrants came from Austria, Belgium, Bulgaria, England, Greece, Holland, Hungary, Ireland, Macedonia, Poland, Russia, Scotland, Switzerland, and Yugoslavia. By far, the largest immigrant population consisted of Italians, and these families developed their own community between Stack and Smith Hills.[40]

The most important development in the Blocton area occurred about one-half mile west of the post office. Settled initially as a business and residential community adjacent to Blocton, Suttletown gained its name from one of its cofounders, Julian LaFayette (Fate) Suttle. Forming a partnership with E. N. Cottingham in 1884, Suttle opened a store to sell general merchandise and mining supplies. Other business interests followed and, by 1890, nearly two dozen stores lined the main street. Incorporated as West Blocton in 1901, this business community became the primary shopping center for residents of Marvel, Coleanor, Piper, Belle Ellen, Lucile, and Moffat.[41]

The fragmented nature of the Blocton-area settlements belied the common tie of coal mining among all residents. Nevertheless, segregation became part of everyday life. In his general study of Appalachia, Shifflett observes that "there is little evidence that mine work mitigated social prejudice against immigrants and blacks." As a rule, black miners received equal pay for equal work; there was no discrimination below ground. However, once outside the mines and above ground, black workers and their families faced separate schools, churches, lodge halls, and amusement centers. Wayne Flynt reveals similar trends in his study of coal mining in Alabama: "Black miners earned the same wage, worked the same hours, ran the same risks, traded at the same commissary, and belonged to the same local union; but they worshiped at different churches, attended their own dances, and were believed to be more superstitious." Blacks also comprised approximately one-half of all coal miners in Alabama. The 1890 census indicates that blacks made up 46.2 percent of the mining population. Native-born white miners comprised 34.9 percent, and 18.7 percent were immigrants. Four decades later, the percentage of black miners expanded to 53.2, whereas native-born whites increased and immigrants decreased to 45.3 and 1.5, respectively. Even so, drinking rooms in the Smith Hill community of Blocton delineated three distinct compartments—"one for whites, another for colored people and a third for dagoes."[42]

In spite of segregated living areas, immigrants generally assimilated well into the coal town social order. As early as 1890, one Blocton resident stated that "all nations work side by side in perfect harmony. . . . The foreigners are especially generous and kind, and few of them would be willing to give up American freedom for their native homes." Another report in 1913 observed that Italian residents in Blocton seemed to relate well within the established social structure. As a rule, European immigrants adjusted relatively easily when they arrived as individuals or in small family units. On the other hand, large groups of immigrant miners recruited by labor brokers fostered resentment. Native-born workers argued that "they are not going to take our places in the mines. We came here first and propose to stay." In Blocton and West Blocton, this nativist reaction to economic competition found voice in the Ku Klux Klan during the mid-1920s. The local organization left no records of its activities, but personal accounts attest to Klan participation in harassing immigrants involved in gambling or in the manufacture of illegal liquor.[43]

In general, European men immigrated to the American coal fields before their families did. At Blocton, these miners lived in the stockade or "Dog Pens" in Number 3 Town until they could repay the cost of their transit. Once established, they sent for their families who arrived speaking little or no English. For example, Rosa Bodnar arrived in the United States wearing a sign around her neck that directed, "Deliver to Charles Bodnar, Blocton, Alabama, U.S.A." Immigrants from other regions assimilated as best they could. Merchants Weinstein and Israel opened the first Jewish-owned department store in Blocton in 1892, promising "all dry goods cheaper than can be bought in New York." Two Swiss families lived within sight of each other at Piper, and they used their traditional yodeling to exchange greetings each morning. The Fullman family moved several times—to Piper (1904), Marvel (1907), Coleanor (1917), Illinois (1918)—and finally settled in Piper around 1920. Also, the Samsal family immigrated from Yugoslavia to Piper in 1930, but later moved to West Blocton.[44]

Even with successful assimilation and general acceptance, the diversity of im-

Fig. 19. "Little Italy," Blocton, Alabama. (Photograph by John T. Weeks. Courtesy of Beverly Weeks Beasley.)

migrants generated a degree of native-born resistance. Segregated from residential areas and ostracized from the fraternal organizations of Blocton, Italian miners formed their own distinctive community in a hollow between Smith and Stack Hills. Situated near Number 3 Mine, Little Italy or "Dago Town" emerged as a Roman Catholic enclave within the Blocton area. These Italian miners constructed their houses along the steep hillsides, often building one dwelling directly on top of another to maximize the use of space and to provide shelter for each family's cows and goats. Corsican native Narcisso Nucci opened a general store, his wife operated a boarding house, and the couple shared an outdoor oven for baking bread. In addition to the communal oven, the Italian residents patronized their own stores, boarding houses, and restaurant. Community activities included a social and literary club known as the Prosperity Society of Blocton. The Yolanda women's club and the Sons of Italy men's club offered social outlets, and a neighborhood boccie court (an Italian form of lawn bowling) preserved an important ethnic custom.[45]

Shifflett considers World War I the catalyst that prompted many European immigrants to leave the American coal fields and return to their native lands. Concurrently, wartime demands for coal created an urgent need for miners. With foreign labor diminishing, operators relied more on native-born white workers. Moreover, as conscription took many single laborers away from the mines, married men with children migrated to the coal towns. This general trend manifested itself in various ways within the Blocton community. As historian Rhoda Ellison explains: "More than the telephone, electric power or the automobile, it [World War I] catapulted the [Bibb County residents] into the twentieth century and opened their provincial eyes to the outside world."[46]

Blocton's population swelled to 5,000 during World War I, and the years from 1914 to 1919 marked record growth for that area. TCI operated Blocton Mines Number 9 and 10 in an effort to meet wartime demands. A troop from Alabama's First Cavalry guarded the mines against saboteurs, and more than thirty Blocton residents enlisted in the Army after the U.S. declaration of war in 1917. In addition, the local chapter of the United Daughters of the Confederacy organized a drive to collect sheets, pillowcases, and tablecloths to send abroad as bandages. Overwhelming support for the war effort included Blocton residents pledging more than $100,000 in Liberty Loan Drives.[47]

Overall, the war had a positive impact on Blocton and other areas within the Cahaba coal field and marked a fitting culmination to the original phase of Cahaba's development. Numerous mining operations provided coal for the nation's war effort, and company towns offered many opportunities for their inhabitants. However, residents also experienced some drawbacks and hardships. For example, Number 3 Mine closed, and Italian miners who lived in Little Italy dug coal in the newly opened Numbers 9 and 10. Mine operators also received a poorer quality of black powder, and miners complained that fumes from the inferior explosives made them sick. Of even greater significance, Ellison notes that "during World War I, the Lucile and Belle Ellen mines depended more than ever on . . . readily available [convict] labor." This source of cheap labor, initiated in Alabama in 1872 and introduced into the Cahaba coal field at Helena's Eureka Mines, brought a dynamic to coal mining that would adversely affect the industry for more than half a century.[48]

5
CONVICT LEASING
1872–1927

On Saturday, April 8, 1911, an explosion rocked the Banner Mine (located in Alabama's Warrior field) of the Pratt Consolidated Coal Company killing 128 miners. Disasters of this type were not uncommon given the risks inherent in such underground work, but the victims' demographics marked this tragedy as one of distinct significance. First, the miners were state and county convicts, leased to Pratt Consolidated. Second, of the 128 killed (122 convicts, 6 free), all but 5 were black. Next, seventy-two convicts (or 56 percent) were from Jefferson County. Finally, 30 percent of the Jefferson County prisoners were serving sentences of no more than twenty days for misdemeanors such as gambling, vagrancy, and illegal drinking. Consequently, the Banner Mine explosion incited the ire of numerous progressive reformers, but opposition to the convict-lease system in Alabama would not end the program until 1927.[1]

The Banner Mine incident constitutes the most infamous tragedy of Alabama's coal industry. In addition, because of its visibility, this disaster drew attention to the paucity of mine safety procedures during the nineteenth and early twentieth centuries. In fact, similar events occurred throughout Alabama's coal district. Just one year prior to the Banner Mine explosion, tragedy struck in the Cahaba field at Lucile shortly after convicts arrived at that location. At 2:00 A.M. on May 16, 1910, fire broke out in the wooden stockade that housed the prisoners. Apparently, three inmates set the fire as a diversion to facilitate their escape attempt. Unfortunately, the flames spread quickly and soon created a life-threatening situation. Many convicts were trapped inside the burning structure, and Warden A. O. Thompson was overcome by smoke as he attempted to unlock cells on the second floor. In the end, the escape proved unsuccessful and an inmate pulled Thompson to safety, but twenty-seven prisoners died, and twenty-two others suffered from burns. Survivors occupied temporary quarters in the commissary building, construction crews rebuilt the stockade, and convict labor continued at Lucile until 1918.[2]

Even though convict leasing failed to gain widespread acceptance throughout the Cahaba coal field, this exploitive system did exert some influence over the region. Adopted at Aldrich, Belle Ellen, Helena, and Lucile, the convict-lease system introduced a statewide issue into the local coal industry. Pitting coerced labor against free miners, the employment of convicts fostered a dynamic of oppression, greed, and frustration in the Cahaba field. Moreover, because of the system's prevalence in the Birmingham District, many operators and miners found themselves vicariously connected to events outside of their local districts. The incidents mentioned are typical of mining conditions throughout the South and reflect the region's unique system for managing prisoners. Similar events occurred at sugar and cotton plantations, turpentine farms, phosphate beds, brickyards, and sawmills that used convict labor.[3]

Historian Matthew J. Mancini contends that the convict-lease system depicted the true nature of the values of the post–Civil War South—racism, violence, shortage of capital, and challenges to modernization. Moreover, convict leasing replaced slavery as the means of controlling the black population. However, Mancini argues that the convict-lease system was "worse than slavery." As property, slaves had value, but convicts were expendable. If a convict laborer died, escaped, was injured or abused, he could be replaced cheaply. Labor historian Alex Lichtenstein agrees that the penal system controlled black labor. But, "far from representing a lag in southern modernity, convict labor was a central component in the region's modernization." Grounded in the heritage of slavery, the convict-lease system promoted economic growth in industry, not in agriculture. In Alabama, Georgia, Tennessee, and other southern states, "Big Mules"—industrialists and planters—combined forces to develop a system of labor recruitment, control, and exploitation that blended tradition with modernization.[4]

Alabama's convict-lease system was the South's longest running one. Noted for its persistence and its profitability, Alabama's lease system traced its roots to antebellum days. State lawmakers established a penitentiary in 1839 and constructed a facility the following year, but the facility ran up a significant debt by 1845. As historian Elizabeth Boner Clark observes, "penology in Alabama labored under the theory that the state's correctional institutions should be self-sufficient." Therefore, the general assembly voted to lease the penitentiary to private interests. Under this arrangement, the state negotiated a contract with a lessee who agreed to pay a specified amount in exchange for the convict labor. Once established, the agreement required the lessee to provide housing, food, clothing, and security for the convicts. In addition, contractors usually agreed to provide transportation from the state penitentiary or county jails as well as a new set of clothes and a ticket home after discharge.[5]

Signing a six-year lease in 1846, J. G. Graham assumed total control of the state prisoners at no cost to the government. Subsequent leases to Dr. M. G. Moore in 1852 and to Dr. Ambrose Burrows in 1858 continued the trend, but netted no real financial gain for the state. Burrows's murder at the hands of an axe-wielding inmate in 1862 prompted a brief period of direct state administration that extended through the Civil War. During that time, convicts manufactured artillery and wagon harnesses for the Confederate war effort. In 1866, the Smith and McMillan subsidiary of the Alabama & Chattanooga Railroad leased prisoners for building new rail lines. The lack of revenue generated by this arrangement caused state lawmakers to reconsider and revamp the convict-lease system. Rather than contracting for the administration of the state prison, legislators opted to lease the prisoners themselves to private enterprises. Thus, by 1872, the formal convict-lease system that would persist until 1927 was in place.[6]

Unfortunately, a revised system did not reverse the trend in death rates among leased convicts that rose from 18 percent in 1868 to 41 percent in 1870. Rather, state officials continually sought relief from overcrowded prisons, idle prisoners, an empty treasury, and heavy taxation. In similar fashion, mine owners sought a more efficient, less expensive workforce. According to most operators, "the convict accomplished more labor than the free laborer, and . . . he is more reliable." When faced with turnover and instability, owners relied on convicts to provide steady, predictable, and controllable labor. In effect, convicts comprised the ideal captive workforce. Representing the least expensive form of labor, prisoners remained generally passive and unorganized. Consequently, when confronted with organized protests instigated by frustrated free miners, operators frequently used their leased convicts as strikebreakers to maintain consistent production.[7]

Milfred Fierce has argued that humanitarian concerns played no role in this system contrived by state officials seeking to minimize penal costs and by lessees who pursued profit as their sole objective. Indeed he described the convict-lease system as a return to slavery. Focusing on similar techniques of race control inherent in antebellum slavery and postbellum convict leasing, he asserts that victimization of blacks formed the centerpiece of the convict-lease system. Viewed by whites as inferior beings and as habitual criminals, blacks remained generally illiterate, unskilled, landless, and ill-equipped to deal with complex political, economic, and social issues. In sum, "convict leasing was indivisible with Blackness." Therefore, labor arrangements that incorporated long hours, substandard conditions, nonexistent medical and health care, poor nutrition, inadequate clothing, oppressive worker-supervisor relationships, punishment akin to torture, and general exploitation closely resembled slavery.[8]

According to Fierce, "sanitary conditions were Neanderthal, living and sleeping arrangements were vulgar, health and medical care were hostage to lessee frugality, and food—such as it was—was frequently unfit for human consumption." Bolstering his argument by quoting noted social reformers, Fierce uses Julia Tutwiler's words to argue that the convict-lease system "combined all of the evils of slavery without one of its ameliorating features." An analysis by Mary Church Terrell also considers leasing a modern regime of slavery: "The convict lease system [was] 'less humane' than slavery because slave owners had an economic stake in keeping slaves healthy and at maximum productive capacity, while no such consideration existed among lessees, who obtained their convict labor by virtue of being the highest bidder at auction." Finally, W. E. B. DuBois contends that leasing "had the worst aspects of slavery without any of its redeeming features. . . . It linked crime and slavery indissolubly in their minds as simply forms of the White man's oppression." These descriptions gain credibility, as, by 1877, blacks comprised 91 percent of the state convict population.[9]

The preponderance of black convicts may be attributed to the effectiveness of the post–Civil War Black Codes. These laws, designed to exert social control over free blacks, limited their freedom to bear arms, keep late hours, or talk with white women. Consequently, many newly freed blacks faced arrest on charges of abusive or obscene language, adultery, gambling, vagrancy, or riding a freight train without a ticket. Mary Ellen Curtin contends that, in spite of ratification of the Fourteenth Amendment and revocation of the Black Codes, state and local laws continued to exert a form of racial repression throughout the Reconstruction period. A two-tiered penal system emerged with convicted felons sentenced to the state prison and those found guilty of misdemeanors housed in county jails. Many county judicial systems required convicts to pay all court costs to include fees charged by the sheriff, the jurors, and the judge. On average, these expenses totaled approximately $50, but "cash-poor sharecroppers" could not afford to pay. Therefore, county convicts remained incarcerated until they could work off their debts at the standard rate of 30 cents per day. As a result, a two- to four-week sentence could be extended to a maximum of eight months at hard labor. Sheriffs, deputies, and court officials also often subsidized their meager incomes with their portions of the fees, so multiple and frequent arrests became common. Many counties limited a prisoner's diet to cornmeal mush and fatback, thereby pocketing up to two-thirds of maintenance stipends granted by the state. By maximizing arrests and minimizing expenses, some local cabals netted nearly $50,000 annually from the fee system. Ultimately, fifty-one of Alabama's sixty-seven counties leased their prisoners to business interests thereby ensuring a large convict-miner population.[10]

Beginning in 1872, the inspector of the Alabama Penitentiary filed an annual report. Initially, these records listed the following data for each convict: name, when received, offenses, term of sentence, age, height, previous occupation, nativity, where convicted, race/gender, conduct, and general remarks (i.e., died, pardoned, discharged, escaped). In addition, the Report of Physician recorded ailments treated and deaths that occurred. The 1874 report included a financial statement that reported the "amount received and due from hire of convicts."[11]

The 1873 report provided a detailed account of leased convicts. For example,

Thomas Williams leased 103 convicts for labor on the state farm in Montgomery County as well as the penitentiary farm in Elmore County. These convicts generated $5,212.50 for the state treasury. Other lease agreements are detailed in table 1.

Table 1. Convict-Lease Contracts, 1873

Lessee	Number of Convicts	Total Amount
Dr. M. G. Moore	6	$250.00
Storrs & Parker (Elmore County)	18	$1,008.00
Boyle & Pollard (Pike Road, Montgomery County)	52	$1,875.00
Alabama Furnace (Alabama Iron Company, Talladega County)	34	$520.00
D. J. Boazman	8	$150.00
New Castle Coal & Iron Co. (Jefferson County)	32	$932.50

These contracts generated total revenues of $9,948. Additional revenues came from leases to Jackson, Morris & Co. of Clanton in Chilton County (17 convicts); Z. P. Crawley (9); Alexander Nummy (1); L. Willis (1); Gaius Whitfield of Linden in Marengo County (1); and W. D. Goggins (1). Thus, of 368 state prisoners in 1875, 283 were leased while only 85 remained inside the penitentiary walls.[12]

Other contractors joined the list of convict lessees in 1876, including the Eureka Company of Helena in Shelby County, Monroe Parker of Equality in Coosa County, Farris & McCurdy of Lowndesboro in Lowndes County, and B. S. Smith of Dadeville in Tallapoosa County. In addition to the expansion of the convict-lease system, the state inspector initiated unannounced quarterly visits to the lessees to inspect convict working and living conditions. He also noted: "Before the passage of an act approved February 10, 1876, the [Penitentiary] Warden was required by law to furnish an officer, at the expense of the State, to attend each squad of convicts so hired out." Commissioned by the state assembly, these inspectors checked on food, living quarters, hospital facilities, and the quantity and quality of clothes. Conducting private interviews with selected prisoners, the inspectors asked the following battery of questions:

1. What do you get to eat, and how often? Is it enough and wholesome?
2. How often do you wash and change clothing? Do you have shoes, hats, and clothing enough to keep you comfortable?
3. Have you been punished? How often? For what offense, and what kind of punishment was inflicted?

4. Have you been sick? How long were you sick? What was the nature of your disease? What physician attended you?
5. At what time do you commence work in the morning? What recess do you have at noon? At what time do you stop work in the evening?
6. Have you any other complaints to render?

In 1877, fourteen lessees controlled a total of 557 convicts who generated revenue for the state in the amount of $26,471.18. The convict-lease system was indeed becoming "big business" in Alabama.[13]

By 1880, state administrators classified convicts according to their physical condition and support requirements. For example, first-class laborers—those who were physically fit and capable of a full day's work—leased for $5 per month. Second-class miners brought in $2.50 per month, and third-class workers were considered "dead heads." This classification was a carryover from slavery days, but a fourth-class designation developed in later years. Sloss-Sheffield records list laborers in the latter category as cooks, farmers, waiters, bath house attendants, hospital orderlies, bakers, barbers, plumbers, wagoners, wheel house workers, and yard men. Even so, workers from all four classifications mined coal. All labored under a task system, which required production of four tons per day for first-class miners, three tons for second-class, two tons for third-class, and one ton for fourth-class. These daily coal production quotas translated into convict miners shoveling eight thousand to twelve thousand pounds of rock and ore six days per week.[14]

By 1880, the list of lessees had changed as well. Comer & McCurdy bought the Eureka Company and administered the Helena mines formerly owned by Henry F. DeBardeleben. Letters from convicts working at Helena described the Eureka mines as "replete with dangers: crumbling walls, gas-fueled fires and explosions, thick dust, and waist-high water." These prisoner accounts also recorded common living conditions: "The rooms were 'filled with filth and vermin.' Gunpowder cans were used for slop jars, chained men suffered miserably when the cans 'would fill up and runover on bed.' On a typical workday prisoners left the camp at three in the morning, in chains, and ran three miles to the mine. They returned at eight o'clock in the evening and stood for hours in the rain or snow to be counted before they could eat."[15]

Forced to exist in this horrific environment, many convicts attempted escape. Proving themselves "difficult to control," these prisoners "made coal companies liable to public exposure, scandal and criticism." In addition, mine operators like J. W. Comer, brother of future governor Braxton Bragg Comer and owner of the Eureka mines, became notorious for their harsh and inhumane treatment of convicts. A firsthand account by a resident of Helena recounts a manhunt that ensued when two prisoners escaped. Apparently, one of the men made good his getaway, but the other fell victim to Comer's brand of punishment. Driven to the ground, the prisoner begged Comer to call off his dogs. Ignoring the man's pleas, Comer "took a stirrup strap, doubled it and wet it, stripped him naked, bucked him, and whipped him—unmercifully whipped him, over half an hour." With the prisoner helpless and begging to be killed, Comer's men "left him in a . . . cabin where . . . he died within a few hours." No criminal charges were filed in this case.[16]

Similar incidents prompted the *Huntsville Gazette* to state that "our present convict system is a blot upon the civilization of this century and a shame upon the State of Alabama. . . . Humanity demands legislation on this subject. Christianity demands it. The State will never prosper as long as the infamous system prevails in her borders." Sadly, such editorials fell on deaf ears, as no legislation limiting the control exercised by mine owners was passed. Rather, coal operators conducted themselves as they saw fit. In addition to his partnership with McCurdy, Comer held an individual interest in the Pratt Mines (owned initially by DeBardeleben), and John T. Milner, former chief engineer for the South & North Alabama Railroad, owned the Newcastle Coal Company. Accusations of inhumane treatment of convict miners at the latter enterprise would become the center of controversy a few years later.[17]

When Rufus W. Cobb (1878–1882) won the gubernatorial election of 1878, he inherited John G. Bass as warden of the state penitentiary system. Under Bass's leadership, the state's penal system generated money for the treasury, but he often resorted to questionable tactics in doing so. Viewing Bass as an unnecessary remnant of George S. Houston's administration, Cobb immediately removed him from the contract bidding process. Realizing significant increases in state revenues thereafter, the governor declined to renew Bass's appointment. Enjoying an improvement in state finances yet realizing the need for penal reforms, Cobb selected John H. Bankhead to succeed Bass. As warden, Bankhead commissioned Dr. John Brown Gaston, president of the Alabama Medical Association, and Dr. Jerome Cochran, state health officer, to inspect conditions for convicts working in coal mines. Gaston and Cochran discovered generally unhealthy and hazardous conditions, and they reported their findings to Bankhead.[18]

Addressing the annual meeting of the state medical association in Mobile in 1882, Gaston criticized the convict-lease system and the contractors for atrocious sanitation practices and horrid living and working conditions. The specific allegation directed against the Newcastle mines sparked a passionate, twofold rebuttal from Milner in which he defended the state system in general and his mining operation in particular. Nevertheless, Gaston's comments prompted a reevaluation of the convict-lease system. In that same year, Bankhead initiated prison reform. He proposed legislation requiring severe penalties for mistreatment of prisoners as well as mandatory periodic inspections of mining conditions—"if practicable." Apparently, the practicability of reform paled when compared with the protection of some $50,000 in annual state revenues.[19]

On the other hand, changes following from Gaston's report and Bankhead's initiatives ushered in more than a decade of penal revision. Under Bankhead's direction, reform legislation constituted a "revolution" in Alabama's prison system. As a result, coal mining emerged as the primary prison industry in spite of protests from agricultural, road-building, and timber interests. As Pratt Coal and Coke Company took control of the Eureka Mines at Helena, Bankhead arranged a system of inspection and regulation. Appointed to Alabama's first Board of Inspectors in 1883, Reginald H. Dawson, Albert T. Henley, and William D. Lee visited mining operations, inspected various aspects of prison life, operated within the state political system, and challenged coal operators. Henley, a physician from Marengo County, and Lee, a Perry County lawyer, served as assistants to Dawson who had practiced law in Dallas County prior to his appointment.[20]

One of the early inspections investigated the Comer & McCurdy operation at Helena. In May 1883, approximately two hundred state convicts worked in the old Eureka Mines. The inspectors' report described the setting as well as the conditions. The mine operators had constructed a two-story, T-shaped wooden dormitory building for prisoners that included a dining hall, chapel, tailor shop, and warden's office. The prison barracks contained four cells (seventy feet by twenty feet) and a dining room opening into a hallway. Cells contained privies, but the inspectors noted their poor construction and offensive odor. A large room above the hallway served as a hospital, but the number of sick prisoners overwhelmed its efficiency and comfort. Overcrowding constituted a problem in the cells as well. Dawson and his assistants reported:

> The men slept upon scaffolds built along the walls; there were two of these scaffolds, and in some rooms three, one above the other, upon which the convicts lay thickly packed at night. There were some old mattresses upon these scaffolds, and there were blankets enough to keep the men warm; but they were all as dirty and filthy as coal-dust and grease could make them, and were thickly infested with vermin. In the four cells there was not more than room enough for the State convicts, and, as there were more county than State convicts at the prison, they were crowded entirely beyond their capacity. There was no ventilation, except through cracks in the wall, which were not always kept open; and as there was a high stockade close to the cells, the circulation of the air was very imperfect. At night, the heat, the stench from the privies, and the effluvia from the persons of the prisoners and from the filthy bed-clothing, made the cells almost unendurable.

Apparently the report generated improvements; by 1884, these same mines boasted a "new dining room, relief from overcrowding, increased supply of water, new hospital, ventilation windows, more light, hanging bunks/new mattresses, renovated bath house, [and] improved latrines."[21]

Upon Bankhead's resignation in 1885, the state eliminated the office of warden and elevated Dawson to chief inspector. Thereafter, Dawson, Henley, and Lee comprised the Board of Inspectors for the Alabama Department of Corrections. Contrary to many of Bankhead's political machinations, Dawson attempted to apply a humanitarian approach to prison management. Based on his personal distaste for the prison mining system, he expressed concern for the welfare of individual miners. Curtin contends that "Dawson often provided a moral voice in the midst of an immoral system." This new approach proved effective, and the trend toward reform continued. For example, a quarterly report submitted in the spring of 1886 concerning the Comer & McCurdy operation at Helena stated: "The convicts have been well fed, very well clothed and kindly treated. The health of the men has been generally very good. The hospital is sufficiently commodious, and kept in a cleanly condition, and the sick received good medical attention."[22]

In March 1886, Dawson recommended and Governor Edward A. O'Neal (1882–1886) approved a new set of "Rules and Regulations for the Government of the Convict System of Alabama, adopted by The Board of Inspectors of Convicts." This list of thirteen rules established standards for ventilation and drainage and required a minimum of two mine entrances to enhance miner safety. The new regulations also stipulated that a mine foreman conduct periodic inspections to ensure compliance with revised safety procedures. In addition, Dawson's requirements promoted efficient mining operations by compelling operators to provide proper tools and equipment to their convict miners.[23]

Dawson's reform efforts centered around "short time," regular releases, letter writing, and family visits. In an attempt to reduce the number of escape attempts, he implemented short-time pardons. Realizing that state law rewarded good conduct by deducting two months from every year of a prisoner's sentence, Dawson established the "time card." Under this system, each prisoner received a time card that recorded his date of conviction as well as his date of release. Dawson modified the latter part by calculating both a short-time and a long-time release date. This practical method for eliminating extensions of regular release dates improved convict morale by encouraging them to "embrace hope and reject restlessness." Dawson also promoted family connections by initiating a formal system of letter writing. Twice each month, the Board of Inspectors provided paper, envelopes, and stamps for the convicts to communicate with their family members.[24]

As Dawson and his assistants instituted their reforms, many prisoners began to exercise a degree of self-determination. Recognizing their skills as miners, convicts began to view themselves as bona fide workers who could effect change in overall mining operations. Exhibiting many of the attributes of free miners and self-employed workers, these prisoners attempted to influence their work environment. Many joined the Knights of Labor and the Greenback Party in an attempt to organize a collective effort for reform. In light of these developments, the Pratt Coal and Coke Company paid overtime wages to its prisoners who exceeded their coal-mining quotas. In turn, this money enhanced self-respect among the convicts; additional earnings provided purpose, fostered identity as a provider, and promoted family connections.[25]

Dawson's reforms did not last, however. In spite of improvements to mining conditions, operators continued to exert authority over the pace of work, punishment, and overall treatment. When the Tennessee Coal, Iron, and Railroad Company (TCI), purchased the Pratt Coal and Coke Company in 1886, convict miners experienced a loss of freedom, work speedups, dangerous conditions, and brutal treatment. When contracts with Pratt and Comer & McCurdy expired on December 31, 1887, the state advertised and solicited bids for new lease agreements. At that time, TCI negotiated an unprecedented ten-year contract for all state prisoners and for half of the county convicts. The remaining county prisoners went to the Sloss Iron and Steel Company despite protests from competing companies. A legislative investigation revealed significant irregularities in the negotiating process, but Governor Thomas Seay (1886–1890) supported the TCI contract. Subsequently, TCI increased its payments to the state, but the contract remained loosely enforced. A total of 1,744 state convicts were leased, with 518 going to TCI and 300 designated for Sloss Iron and Steel. In addition to these state prisoners, 622 county convicts were leased, most of these allocated to the Pratt Mines and to Coalburg in Jefferson County.[26]

TCI retained the practice of compensating miners for extra work. In fact, the company paid more than $7,000 to convicts in 1890. Some of this outlay returned to TCI coffers as prisoners purchased cigars, cigarettes, fruits, and canned goods from the company's "convict merchant." After the TCI takeover, convict miners resorted to gambling for sporting entertainment and to spending more money on themselves through internal prison markets. Consequently, fewer dollars went home to provide for the needs of their families. Sloss, too, compensated its prisoners for extra work, but officials insisted on paying with company scrip lest miners earn enough cash to pay their court costs and thereby gain their freedom. In effect, scrip payments guaranteed a captive workforce.

These undercurrents prompted convict miners to band together in protest. Many prisoners used Dawson's writing program to compose letters of complaint to the inspectors and to the governor. Others became more cynical and resorted to direct action such as setting fires, planning escapes, committing suicide, or stopping work. Sabotage, strikes, and other forms of resistance exemplified the miners' frustrations, but they also revealed the prisoners' resolve. As Curtin points out, these forms of protest did not represent random acts of rebellion but deliberate responses designed to evoke specific changes. Growing unrest among Alabama's convict miners indicated the need for further reforms.[27]

On February 14, 1893, the state assembly adopted "An Act to create a new convict system for the State of Alabama, and to provide for the government, discipline and maintenance of all convicts in the State of Alabama." The new law stated that "no convict, State or county, shall be employed in any mines in this State, except as provided for in this act." However, Section 48 continued:

> *Be it further enacted,* That upon the termination of the contract with the Tennessee Coal, Iron and Railroad Company [TCI] . . . said contract shall not be renewed or extended. . . . That it is the true intent and meaning of this act, that the convicts now worked under contract in any of the coal mines of this State shall be removed therefrom as rapidly as practicable, by the first day of January, 1895, *if it can be done without detriment to the financial interests of the State.* [emphasis added]

In spite of support for reform from Governor Thomas G. Jones (1890–1894), the Panic of 1893 thwarted all attempts to remove convicts from the mines. Citing the need to preserve vital sources of state revenue, Governor William C. Oates (1894–1896) convinced the legislature to repeal all legislation introduced by his predecessor. The ineffectiveness of the 1893 decree is evident when contrasted with the implications of a similar act passed in 1895 under Oates's leadership. This act—to regulate the management of state and county convicts by the General Assembly of Alabama—stipulates that "convicts must be classed or tasked, if hired in mines." Furthermore, contracts must specify the type and place of labor; convicts were restricted to that place and to that work unless a new contract were negotiated.[28]

Thus, in spite of Dawson's heroic efforts at reform, the state penal system continued to generate significant revenues for the state. Dawson hoped to alter the county convict system as well, but as chief inspector he continued to enforce current state policy. State officials relied on the system to maintain social order and to produce income. Almost impervious to Dawson's reforms, convict-mining became an institution unto itself. When he retired in 1897, Dawson faced the realities of failure in achieving his ultimate goal—abolishing the lease system and removing the convicts from the mines. His successor, S. B. Trapp, reversed many of Dawson's reform measures in his support of the status quo. He subscribed to the widespread commitment to white supremacy and prison profitability. Consequently, by the turn of the century, Alabama's convict-lease system represented the most profitable such enterprise of all the states.[29]

There was no real effort to terminate the leasing system in the early twentieth century, but some minor reforms did occur. However, early in his administration, Governor William D. Jelks (1901–1907) boasted that, under his leadership, the state penal system earned $100,000 in one year—more than his predecessor, Joseph F. Johnston (1896–1900), netted during his entire four-year term. In similar fashion, Governor Braxton Bragg Comer (1907–1911) attempted to offset budget deficits by promoting convict leasing. Even with this emphasis on profits, token humanitarian policies reduced some of the hardships and corruption within the system. For example, the Rules and Regulations of 1886 were published under separate cover in 1901.[30]

Intrigue and corruption continued to plague the penal system. A new contract with TCI in 1904 transferred responsibility for housing and feeding prisoners from the company to the state. In return, TCI agreed to pay a fixed price for every ton of coal mined by convict labor. The ensuing increase in state revenues prompted James G. Oakley, president of the Board of Convict Inspectors, to move prisoners from turpentine and lumber camps until almost all state convicts worked in coal mines. Some years later, Oakley's chief clerk, Theophilus Lacy, siphoned $115,000 from the record-breaking revenues of 1913. Arrested the following year and charged with embezzlement and grand larceny, Lacy was sentenced to sixteen years in prison. Similarly, Oakley faced trial twice for embezzlement of state funds, but, unlike Lacy, he was acquitted on both counts.[31]

Increasing revenues reflected expansive growth, and the convict-lease system extended its reach to three mining communities in the Cahaba coal field. A series of quadrennial reports commencing in 1910 trace the growth of Alabama's system. In that year, TCI leased 450 convicts, and Sloss-Sheffield assumed control of 250. Henry F. DeBardeleben's Red Feather Coal Company at Lucile leased two hundred prisoners, and his Bessemer Coal, Iron, & Land Company at Belle Ellen leased an additional two hundred. The 1914 report recorded a total of four hundred convicts at the Pratt Consolidated Coal Company's Banner Mine with total revenue for the state above $300,000. Sloss-Sheffield maintained its workforce of 250 and generated nearly $450,000 for the state treasury. The Red Feather and Bessemer Coal enterprises employed 300 and 250 convict miners, paying $262,000 and $301,000, respectively. TCI, having terminated its contract in 1912, contributed almost $175,000 in additional funds. Finally, the Montevallo Mining Company at Aldrich—a relative newcomer to the convict-leasing system, starting the

Fig. 20. Belle Ellen Prison. (Courtesy of Beverly Weeks Beasley.)

practice after the Aldriches sold the company in 1912—added about $17,000 more. By 1922, revenues had increased to the following amounts:

Montevallo Mining Company	$717,689.84
Bessemer Coal, Iron, & Land	$686,127.17
Pratt Consolidated Coal Company	$848,862.66
Sloss-Sheffield Steel & Iron	$957,581.57

Total revenue generated by these four major lessees topped $3 million for this four-year period.[32]

At the same time, leasing rates increased to challenge the free labor rate of 85.5 cents per ton of coal mined. First-class laborers leased for $93.125/month, second-class for $83.125/month, third-class for $73.125/month, and fourth-class for $63.125/month. In effect, the cost for convict labor was on par with that of free labor, but coal operators opted to maintain the convict-lease system because of its reliability. Moreover, convict miners could not move to a new job, choose to stay home from work, or exercise any degree of freedom concerning their employment.[33]

Advocates of the convict-lease system focused on the economic benefits to the state treasury rather than on the humanitarian needs of the prisoners. Reformer George Washington Cable provided a realistic view of the system when he stated:

Fig. 21. Convicts in Aldrich, Alabama. (Courtesy of Henry Emfinger and Aldrich Coal Mine Museum.)

"without regard to moral or mortal consequences, the penitentiary whose annual report shows the largest cash balance paid into the state's treasury is the best penitentiary." Certainly, convict leasing represented an important economic venture, and the system's prosperity kept taxes low for state landholders. Moreover, the following data in table 2 illustrate the growing importance of the convict-lease system within the state's overall fiscal plan. As Clark observes, "The officials of the state hesitated to abolish the lease system that netted the state so great a profit."[34]

Other factors stemmed from convict leasing as well. Ronald Lewis notes that 80 to 90 percent of all convict miners were black. Hence, he argues, as many contemporary reformers did, that the system served to reinstate a form of slavery by establishing a racial hierarchy. Furthermore, because law enforcement and court

Table 2. Economic Benefits of Convict Leasing

Revenues			
Year	State	Convict Dept.	Convict Lease as Percentage of State Budget
1914	$6,607,001	$1,162,493	17
1919	$8,558,751	$1,666,089	19
1923	$18,692,362	$2,629,696	15

Convict-Lease Profits for Alabama Treasury			
Dates	Total Profits from Convict Leasing	Profits from Convict Miners	Mining Profits as percentage of Convict Leasing Revenues
1910–1914	$2,188,604	$1,325,182	60
1914–1918	$2,635,686	$2,059,963	78
1919–1922	$3,671,210	$3,357,354	91
1922–1926	$3,269,098	$2,590,533	79

officials were paid by commission rather than on salary, sentences even for misdemeanors often increased up to three times. Lewis also argues that the convict-lease system promoted political bargaining within the restored Democratic Party. An acceptable compromise developed between Black Belt planters, who advocated social control, and capitalists of the Birmingham District, who promoted industrial progress and growth. These "Redeemers" and proponents of the New South defended the system as a means of rehabilitation and technical or vocational training. They justified this position by arguing that more than half of the free miners in the Birmingham District learned their trade as convicts. Entrepreneurs also supported the state system because of its consistent production, reliable workforce, cheap labor, antiunion sentiment, and source of ready-made strike breakers.[35]

On the other hand, a death rate of 9 to 10 percent in Alabama mining camps did not compare favorably to the 1 to 2 percent rate in the mining areas of Pennsylvania and Ohio. Reformers attributed death and disaster to poor sanitation and harsh working conditions. As in the Banner Mine tragedy, the death of convicts who had committed relatively minor crimes and received comparatively short sentences constituted "judicial murder." In 1913, organized demands for the abolition of the convict-lease system gained momentum with the formation of the Alabama Convict Improvement Association in Birmingham. This group advocated removing convicts from the coal mines and employing them on the roads of Alabama. In the same year, the U.S. Good Roads Congress met at St. Louis and resolved to encourage use of convicts on public highways. This action, according to the resolution, would promote "better morals, reformation of convicts, improved roads, and a higher order of intelligent citizenship."[36]

Similarly, a report from Alabama's Legislative Investigating Committee, submitted on July 13, 1915, criticized the state's penal system:

> The convict lease system of Alabama is a relic of barbarism, a species of human slavery, a crime against humanity.... We find under this driving slavery system, [that] the free miner mines two tons, the convict produces four.... It has been too much the policy of this state to look upon the commercial side of convict life. Each successive administration has done all in its power to increase the earnings of the convict department. The humanitarian side of the question has been entirely lost sight of.... The average life of a convict sentenced to work in the mines is seven years. The effect of the system is to make by the process of death, long-term convicts into short-term ones.... We therefore find the convict coal mines operating on full time, at full capacity, with the convict driven to the task of from ten to twelve [hours] a day, while the mines of free labor employers are operating at a loss only two or three days a week.

In spite of growing support among antilease factions, humanitarian reformers lost momentum with the onset of World War I. As unmarried free miners responded to the draft and as European immigrants returned to their native lands, convict miners represented one of the most stable aspects of the coal-mining industry. In essence, the convict-lease system proved critical in sustaining wartime production levels.[37]

The spirit of reform revived after the war as Governor Thomas E. Kilby (1919–1923) advocated the abolition of the convict-leasing system. His attempts to sway the legislature proved futile, however, as lawmakers recognized the difficulties inherent in replacing a system that generated $750,000 in annual revenues. In 1923, antileasing forces formed the Statewide Campaign Committee for the Abolishment of the Convict Contract System. A year later, the "wrongful death" through water torture of convict-miner James Knox resulted in a public outcry against the lease system. At that time, state officials decided that, instead of leasing convicts to the mines, they would lease the mines and work the prisoners in them. This conversion to state-operated coal mining began at Belle Ellen in February 1924, continued at Flat Top in July 1924, and extended to Banner and Aldrich in March and August 1925, respectively.[38]

Even with the groundswell of opposition against the leasing system, most of the humanitarian reform movements failed to gain any credibility. As Matthew Mancini points out, officials often dismissed them as naïve efforts by "Weeping Willow women and sobbing sissy men." Nevertheless, reformers gained a champion in Bibb Graves who campaigned in 1926 on a platform committed to abolishing the convict-lease system. Before taking office, governor elect Graves (1927–1931, 1935–1939) persuaded outgoing governor William W. Brandon (1923–1927) to convene a special session of the legislature in December 1926. During this session, legislators provided for a constitutional amendment that established a $25 million bond issue for maintaining the state's roads. The new law instituted an additional 2-cent tax on gasoline and created eighteen road camps complete with buildings, road-building machinery, and other equipment.[39]

Only when this revenue-generating alternative arose did state legislators terminate the leasing system and revise the penal system in Alabama. The gasoline tax of 1927 provided alternative funding, and convicts left the mines to work in road gangs and on the state prison farm. Even with its termination on June 30, 1928, the convict lease remained true to form. After an infamous half century of cruelty and corruption, state officials terminated the system based on economic issues rather than on humanitarian concerns. Because of its exploitation of captive labor, the convict-lease system undermined public morality and constructed its notorious legacy. As Isadore Shapiro, president of the Alabama Committee on Prisons, stated in his address to the American Prison Association meeting in New Orleans in 1917: "The damnable system which sells human beings at auction to the highest bidder, . . . Alabama's lease system is her unholiest and most indefensible shame."[40]

6
WELFARE CAPITALISM

1915–1933

On the surface, paternalistic relationships within the towns of the Cahaba coal field appear mutually beneficial. For example, at Aldrich, William F. Aldrich retired in 1905 and sold his interest in the Montevallo Coal and Transportation Company to his brother, Truman. In spite of this change in ownership, William and his wife, Josephine, remained at Rajah Lodge until 1912. In the meantime, Aldrich built Farrington Hall (named for his son) to house his personal library, the company offices, a private club, and the doctor's office. During the seven-year period between retirement and their subsequent move to the Glen Iris neighborhood in Birmingham, William and Josephine "were friends and advisors of all their employees and all who came within their influence."[1]

At Piper, paternalistic relationships seemed to foster goodwill between management and labor. Walter Henley, president of the Little Cahaba Coal Company (LCCC), constructed a swinging bridge across a ravine between Piper and Coleanor to provide a shorter and safer route for schoolchildren. Moreover, Henley, who enjoyed a "father-son" relationship with owner O. H. P. Piper, returned the favor by sending Jess Shepard, son of Henley's cook and housekeeper, to college. Shepard eventually became treasurer of the company. Shepard states that "many people came to see him [Henley] for advice on various subjects, and he always took time to talk to them." In similar fashion, the company secretary-treasurer, Benjamin H. Sherrod, offered Cecil Sewell, a newly ordained minister, a job as his bookkeeper. The older Sherrod and the younger Sewell both taught boys' (ages 9–16) Sunday School—one Methodist, and the other Baptist. Also, Superintendent Percy Randel promoted a wholesome community environment. Sewell claims that Randel developed "good places to live and rear families. Much of the wholesomeness of these communities could be credited to Percy Randel."[2]

Another example of purported mutually beneficial paternalism may be illustrated by Margaret. This small community's locally published history is dedicated "to the DeBardeleben Family who laid the foundation for a legacy that still remains with the people of Margaret today." Almost idolizing founder Henry F. DeBardeleben, local historian Marie Butler records that he was "concerned with the welfare of his fellow man. He had a deep faith in God and a dream that . . . thriving, happy communities could be founded." Even with such accolades, true credit for Margaret's legacy rests with Charles F. DeBardeleben.[3]

The younger DeBardeleben, "Uncle Charlie," inherited the Margaret mines in 1910. By 1914, he had organized Welfare Societies in which employees elected representatives to negotiate issues with the company. According to the stated objectives, these community organizations served "to promote all means and measures that may be of benefit to the employees of the company." DeBardeleben's system created committees to coordinate beautification projects, monitor sanitation procedures, and provide financial relief. In addition, the local societies promoted religious services, education and school maintenance, and recreational activities. Butler reports that "at the monthly Welfare meeting, employees could submit any problem regarding jobs or families with the knowledge that it would be dealt with satisfactorily."[4]

When the Great Depression began in 1929, DeBardeleben took direct action to benefit his employees. Hiring an agriculturist to teach basic gardening skills to his miners, DeBardeleben purchased mules and farm equipment for their use. Providing hogs, chickens, cows, and seeds at cost, he offered prizes to residents who developed productive gardens. As many of the women learned canning and preservation techniques, DeBardeleben encouraged his employees to can at least one hundred quarts of surplus fruits and vegetables. By 1937, approximately four thousand acres of company land was under cultivation, and Alabama Fuel and Iron Company (AF&IC) employees displayed many of their foods at the Alabama State Fair. This practice was not unique to Margaret, Cahaba, or Alabama, however. In his study of West Virginia mining operations, David Alan Corbin argues that gardens and house improvements occupied the miners' spare time and prevented unrest. For minimal cost, operators could upgrade the physical environment, generate community pride, and "stimulate neighborly (social) competition in a rigidly ordered social structure."[5]

In 1932, as effects of the Depression settled on the mining communities, AF&IC employees adopted a resolution pledging their continued loyalty to the company, their support for company policies deemed necessary during hard times, and their willingness to work an extra hour each day to help the company. DeBardeleben responded by erecting a plaque to commemorate the event, and he organized a Quarter Century Club in 1935. This organization consisted of eighty-one charter members, and each miner received a gold pin and an additional $5 per month. An example of this community work ethic occurred on a Sunday when 1,500 AF&IC workers mined four thousand tons of coal to be distributed to needy people in Arkansas, Louisiana, Mississippi, Florida, Georgia, Tennessee, and other parts of Alabama. Also, on one Christmas Eve when a strike directed by the United

Mine Workers of America deprived the Birmingham vicinity of its coal supply, AF&IC miners dug and distributed coal for areas in need of heating fuel. "Uncle Charlie" assured his employees that the coal would be used solely for heating, and he awarded a turkey to the miner who dug the most coal.[6]

This spirit of cooperation seemed to permeate the entire Margaret community. Time and again, DeBardeleben and his employees resisted overtures from the United Mine Workers. As union activity increased in the 1930s, DeBardeleben raised his wages to equal those paid by union mines. He continued his paternalistic approach to leadership, and AF&IC employees remained loyal to him and to the company. An armed confrontation in 1935 brought issues to a head, but Margaret residents remained loyal to DeBardeleben until the mines closed in 1950.[7]

This paternalistic approach to business traced its origins to the outbreak of World War I in August 1914. As the conflict in Europe attracted immigrant workers back to their homelands and as conscription claimed numerous single miners, U.S. industry experienced a significant decline in its labor force. Consequently, industries in the Northeast and the Midwest actively recruited black laborers from the American South. Encouraging African Americans to leave the land of Jim Crow, corporate agents enticed black workers with the promise of regular work, better pay, and greater self-respect. From 1915 to 1930, free black coal miners, along with other southern industrial workers, hearkened to the siren song and joined the Great Migration northward.[8]

Thus, emigration, conscription, and migration changed the demographics of southern coal towns. As skilled African American workers moved northward, other unskilled blacks turned from farming and sharecropping to mining and industry. Furthermore, when single miners enlisted or responded to draft notices, married miners and their families migrated to the mining settlements. As Crandall Shifflett describes, these phenomena represented a fundamental shift in the nature of coal towns and initiated the "Paternalistic Period."[9]

During this phase of paternalism from 1915 to 1930, coal operators focused on the needs of the miners and their families. Noting the change toward a family-based workforce, companies converted some of their assets into schools, churches, recreation centers, libraries, and beauty parlors. As quality of life improved, company towns assumed a community spirit, and the heavy drinking, gambling, and violence of the earlier camps declined. Even though many working-class families adopted middle-class values, Shifflett argues that "mining families were capable of accepting the benefits of paternalism without compromising their independence entirely." Contrary to historians who believe that coal miners fell victim to calculating operators and exercised little or no control over their destinies, Shifflett contends that mine workers maintained a degree of agency in determining their future.[10]

On the other hand, Shifflett admits that the absolute authority of operators manifests itself most clearly in a paternalistic environment. Equating paternalism to "contentment sociology," he sees paternalistic gestures combining aspects of social control, exploitation and oppression, and benevolence. Arguing that labor expediency and financial security, rather than altruism, motivated the mine owners, Shifflett asserts that "operators believed that paternalism was simply another cost of doing business." As the supply of labor dwindled and competition for workers increased, companies frequently compromised established values and proper conventions in an effort to increase production.[11]

According to Corbin, West Virginia coal towns extended their influence beyond welfare capitalism. They undermined militant solidarity, fostered different community values and cultural norms, and dictated both social and economic behaviors. Corbin states: "Plantation slavery was based on color, company-town industrial slavery was based on class." Therefore, any form of kindness, benevolence, or paternalism was designed to make class differences acceptable to the miners.[12]

Stuart Brandes's study of American welfare capitalism supports Shifflett's and Corbin's viewpoints. Brandes attributes the inception of the paternalistic concept to Samuel Slater at Pawtucket, Rhode Island, in 1790. Using Sunday School as a recruitment tool for attracting boys to his textile mill, Slater sought to overcome the challenges inherent in industrialization: mass production, economic growth, immigration, and other societal factors. Company housing, profit sharing, libraries, gardens and beautification projects, schools, and sports and recreational activities all stemmed from the same stimulus. In effect, Brandes contends that "welfare capitalism" is synonymous with "industrial paternalism." Similarly, "industrial village" equates to "company town," and "representation plan" holds the same connotation as "company union."[13]

Welfare capitalism—or paternalism—then, by definition, means "any service provided for the comfort or improvement of employees which was neither a necessity of the industry nor required by law." Some historians assert that the value of welfarism lies in its resistance to trade unionism, and others suggest that paternalism transformed the U.S. corporation from an economic institution into the basic unit of American society. For Brandes, welfare capitalism constitutes "a means of indoctrinating workers into accepting corporations as the central institution of modern American life."[14]

Even before U.S. involvement in World War I, corporations developed programs to "Americanize" their immigrant workers. With no federal support for Americanization, corporations proceeded unilaterally to teach English to their foreign laborers, to encourage immigrants to pursue U.S. citizenship, and to promote American customs and traditions within the workforce. As war clouds loomed, the Wilson administration took notice of welfare capitalism as an effective means for limiting labor turnover, stimulating production, and meeting wartime demands.[15]

Increased labor militancy in the postwar period heightened corporate recognition of paternalism as a counter to trade unionism—"the common devil which brought American businessmen together as no other issue did." In fact, many operators measured the value of welfare programs by their effectiveness in preventing strikes. An executive from Plymouth Cordage stated that "the greatest argument for [welfare work] is that the professional labor leader is opposed to it. He hates it." Moreover, American Federation of Labor president Samuel Gompers referred to welfarism as "Hellfare Work." Thus, welfare capitalism and trade unionism inversely affected life in mining communities. As union activities increased, corporate leaders devoted more time to paternalistic programs. When union organizers appeared less influential, business executives redirected assets to other concerns. As historian Marlene Rikard states: "Welfare capitalism was the employers' alternative to collective bargaining."[16]

Many of these general trends found application in the coal-mining industry, and the "company town" became the central aspect of paternalism. Operators provided company houses, doctors, schools, sports teams, stock, and "welfare societies," but humanitarian concern for the workers did not necessarily define the central motivation for such action. For example, one executive commented: "When I keep a horse and I find him a clean stable and good food I am not doing anything philanthropic for my horse." Others charged for their services and activities to maintain a businesslike approach to welfare capitalism. When measured in "dollars and cents," welfarism followed economic trends. Corporate officials willingly devoted assets to paternalistic programs while business remained strong, but a weaker economy precipitated cutbacks in those same perquisites. Consequently, when the Great Depression ensued with the stock market crash of 1929, financial straits dictated the termination of welfare capitalism.[17]

In sum, paternalism represents "an array of practices designed with specific purposes and carefully considered strategies in mind." Cynical operators often described their laborers as ignorant, slovenly, lazy, drunken, extravagant, and disloyal. These undesirable traits fomented inefficiency, demands for higher wages, labor turnover, militancy, and unionism. In contrast, many business owners sought to create the ideal worker—a laborer identified by thrift, cleanliness, temperance, intelligence, industry, and loyalty. Bolstered by Frederick Winslow Taylor's time-work efficiency studies and by cries from Progressive reformers, American corporate leaders supported welfare capitalism to improve efficiency and to increase production.[18]

Generally, management's paternalistic approach to labor defined the work environment. Operators gained a degree of social control, but workers also enjoyed a better life. In this case, Brandes focuses on the manipulation and exploitation of a passive labor force, but Shifflett identifies opportunities seized by workers in an attempt to affect their environments and to determine their own destinies. In the Cahaba coal field, paternalism manifested itself through the commissary or company store, health care, housing, churches, schools, and recreation and leisure activities. Hence, miners and their families experienced both aspects of welfare capitalism—a dynamic that Piper historian James Walker calls "the struggle and the joy."[19]

In general, historical studies of the company town conjure negative feelings, but, as Shifflett discovered, many oral histories and personal recollections portray a positive image. The U.S. Coal Commission Report of 1925 identified 713 of 880 coal towns as "company-controlled" communities, and the Boone Report of 1946 included photographs of the best and worst coal-mining settlements. Even though Piper ranked among the lowest tier of towns in the former report, its people enjoyed a common bond and a keen sense of civic pride. Thus, Shifflett concludes that economic and social relationships rather than physical features and facilities defined a sense of "community" among mining families.[20]

As a rule, operators positioned the commissary or company store in a central location. At Blocton, the Tennessee Coal, Iron, and Railroad Company (TCI) commissary filled a two-story building at the center of the cluster of settlements. In addition to its prominent geographical placement, Shifflett reports that "the company store was the social and economic nexus of the company town." Generally, the commissary stocked everything that mining families might need. When necessary, the store manager could order out-of-stock items, and the railroad served as the main line of supply. At Piper, the commissary carried dry goods, groceries, meats, hardware, medical supplies, automobile accessories, feeds, toys, and sundry other items. A partition ran the length of the store, separating the dry goods area from the grocery department. A detached powder house stored mine explosives, and a designated clerk dispensed those items to miners on demand. Nearby, the barbershop building included a game room for ping-pong, checkers, and other games. After a fire in 1947, the barbershop shared a building with the dry cleaning plant.[21]

The company store served as a gathering place as well. Impromptu meetings on the commissary porch or around the "pot-bellied" coal stove often reached across established social lines. Residents crossed paths as they went about their daily routines. For example, the Aldrich commissary, constructed in 1923, initially contained the post office, telegraph office, and depot. Later, a two-story building

Fig. 22. Commissary staff in Piper, Alabama: (left to right) Martin Wallace, E. C. Day Sr., Thomas M. Fancher, Herman Jones, Jessie Wells, B. F. Farrah, Murray Langston (manager), H. P. Eady. (Courtesy of the author.)

with full basement housed the company store, but it remained in a key location near the railroad and post office. At Piper, the commissary building included the post office, and residents gathered each morning to check mail, shop, and visit. Similarly, Coleanor combined the store, company offices, and post office in one building so residents could take care of paychecks, money orders, mail, and food at one location. Ordinarily, games of dominoes and cards occupied some porch space while children passed the time playing hide-and-seek, shooting marbles, or spinning tops. In Margaret, a large rock building contained the company store and provided enough room for games as children waited for their parents.[22]

Obviously, the company store served the community as the primary provider of basic services, social interaction, and recreation. As the focal point for the town, the commissary also offered a means for implementing company policy. As Shifflett notes: "Since company paternalism was mediated through the company store, . . . [it] could easily be overlooked by the miners as an instrument of manipulation and social control." In practical terms, the commissary encountered minimal competition. Local stores, vendors, and mail-order businesses provided the only challenge to a virtual monopoly. To counter competition, managers at the LCCC's commissaries in Piper and Coleanor communicated by telephone to manage their inventories and to standardize prices. All in all, Brandes identifies four main complaints against the company store:

1. Prices were supposedly higher than in independent stores.
2. Companies allegedly exerted explicit or implicit pressure on employees to get them to shop in company stores.
3. Company stores were felt to lack variety in kinds of goods carried.
4. Companies were accused of preventing competitors from starting stores nearby.

Throughout the coal-mining industry, company stores were known as "gip-me stores," "pluck-me stores," "gip-joints," and "robber'saries."[23]

Viewing the company store as an instrument of manipulation and social control, Shifflett identifies it as the "official arm of the company." As the financial hub of the coal town, the commissary merely recycled company money and made a profit at the miners' expense. Wayne Flynt agrees; commissaries promoted themselves as the convenient choice, but making a profit for the company remained their primary goal. Typically, commissary prices ranged from 4 to 7 percent higher than those at independent stores, but the company offered delivery and quality assurance. Paid on "bill day"—the fifteenth and the last day of the month—miners used checks or coupon books to place their orders. This two-week projection, known as the "estimate," reflected anticipated needs for the next fortnight. Families gave their lists to the commissary clerks who filled the orders and prepared them for delivery by truck or by mule and wagon. Special orders, such as ice cream for the Fourth of July or toys at Christmas, also originated at the company store.[24]

In many cases, the company office shared a building with the commissary. In effect, miners could receive and spend their wages at the same location. Pay deductions, or "stoppages," paid for rent, lights, hospital insurance, burial funds, schools, churches, tools, black powder and fuses, clothes, and household coal. Furthermore, "scrip" served as company money when miners needed an advance on their next paycheck. "Scrip," "clacker" (taken from the French *claquer,* and based on the noise it made in one's pocket), or "boogaloo" consisted of hard cardboard or plastic tokens. Redeemable only at the commissary, this company money helped to further the store's monopolistic practices. Shifflett's research reveals that 72 percent of black miners spent their earnings at the company store. Native-born white workers spent about half (51 percent) and immigrant families about one-third (33 percent) of their incomes within the company town. Additional data indicate that 50 to 70 percent of all earnings were paid in cash; miners received the balance in scrip.[25]

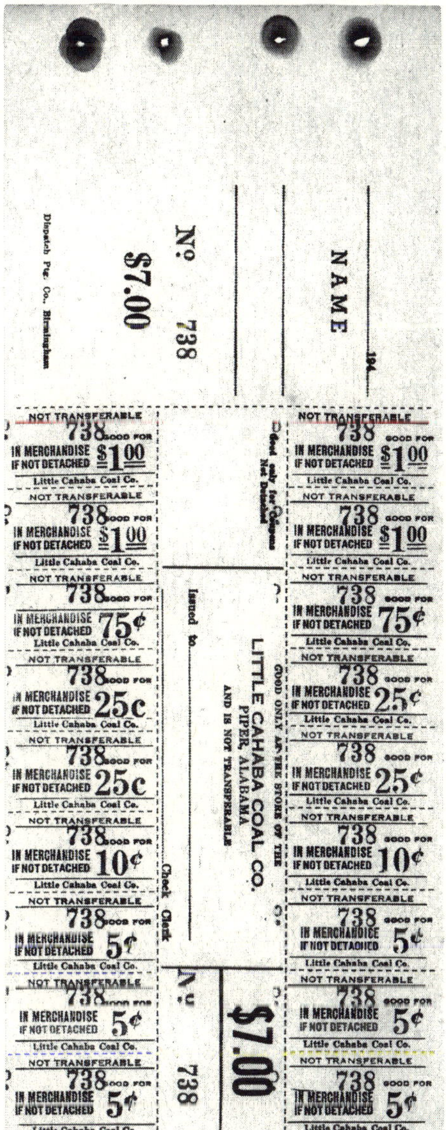

Fig. 23. Scrip from the Little Cahaba Coal Company in Piper, Alabama. (Courtesy of the author.)

The standard exchange rate valued scrip at 80 percent of cash value. Miners who needed an advance on their wages could receive payments in scrip before "bill day." Because these advance payments could be redeemed only at the commissary, the use of scrip usually engendered negative connotations. Miners frequently complained of "outrageous prices, a monopoly on essential food and supplies, and crippling debt." They also criticized the system for its "monopolistic prices," "outrageous deductions," "audacious robbery," and "peonage." In sum, miners "owed their souls to the company store."[26]

On the other hand, some miners used the scrip system to their advantage. Shifflett counters the stereotypical response by indicating that many miners used scrip prudently and that they managed advances and conversions carefully. In Margaret, miners who received advances in clacker often negotiated with fellow miners for converting to cash at the standard 80 percent exchange rate. Another technique for maximizing buying power involved John Gates, a miner at the Paramount Mines in Helena. According to Gates, he used his company scrip to buy snuff at the commissary and then walked to Helena where he exchanged the snuff for groceries.[27]

Some companies offered expanded financial options to their employees. Officials initiated profit-sharing and stock-ownership programs in an attempt to boost production. Also, mutual benefit associations and group insurance plans supplemented the standard check-off or dues system. The former paid death benefits ranging from $10 to $2,000, and the latter paid a $1,000 death benefit or $10 per week (for a maximum of thirteen weeks) in the event of illness or injury. At Blocton, the Cahaba Coal Company provided a "Savings Department" through which workers could invest a minimum of $25 to be managed by various Birmingham-based banks. These deposits earned 4 to 6 percent depending on the length of deposit.[28]

Pension plans also gained popularity among operators and miners alike. Brandes identifies three basic types of plans. Control of the "informal program" rested with the employer who decided eligibility requirements, pay-out amounts, and termination deadlines. In the "formal-discretionary program," the employer retained full authority, but he established standard regulations for eligibility and delegated administrative requirements to a committee of subordinates. Finally, the "limited-contractual program" provided full discretion with no legal obligation to the employer although most operators agreed not to reduce any established pension. According to Brandes, approximately three-fourths of all pension plans adhered to the first two models, and the remaining 25 percent followed the latter program.[29]

Certainly, financial planning constituted a two-edged sword. Companies accepted the expense of such incentives by focusing on the ultimate outcome—

increased production. Also, within the framework of welfare capitalism, these programs served as counters to inefficiency, unionism, and strikes. On the other hand, some form of financial security benefited the workers as well. Assured that their families would be all right in the event of death or injury, miners developed a sense of company loyalty that translated into a stable workforce.[30]

Brandes observes that the concept of a company doctor gained acceptance as interest in the mutual benefit associations and group insurance plans increased. Although reliance on company health services might compromise miners' sense of independence, the quality of company-sponsored medical care usually exceeded that available to individual workers. From the company's perspective, applied medicine would benefit miners and their families directly while improving labor stability and production levels. In 1907, under the leadership of George Gordon Crawford, TCI initiated its welfare system to address social programs, education, and housing. In many ways, Crawford's plan adhered to the standard model for welfare capitalism, but Rikard asserts that family health care formed the "fulcrum" of TCI's system. Corporate research indicated that high turnover rates and absenteeism did not necessarily stem from idleness and transiency. Rather, health and environmental problems such as disease, polluted water supplies, and poor nutrition adversely affected worker performance. In fact, in spite of widespread publicity and interest in mine accidents and explosions, more deaths occurred due to disease and illness caused by unsanitary conditions.[31]

Crawford believed that education represented the critical element in generating proper health, hygiene, and sanitation measures. Consequently, doctors confronted small pox, typhoid fever, diphtheria, tuberculosis, hookworm, venereal disease, and pellagra. Furthermore, health and sanitation specialists faced challenges such as redesigning, cleaning, and maintaining outdoor privies, vaccinations, bath houses, and health education programs. Health officials established methods for eliminating standing water, checking stables, disposing of garbage, and improving water supplies. Based on the premise that sanitary habits equated to mine safety, doctors inspected milk deliveries and ensured that commissary workers handled and safeguarded food properly. Nurses instructed mothers in prenatal and postnatal care, nutrition, and housekeeping. Daughters received instruction in basic sex hygiene and in the "fundamentals of eugenics."[32]

Social workers taught wives and daughters how to care for babies, prepare foods, and sew clothes. Focusing on children, young people, and women, social workers played a prominent role in the TCI system. Hiring the first full-time social workers in 1911, Crawford appointed Marian Whidden to plan and direct TCI's program. Whidden expressed her vision as follows: "Give the children a chance. They will become the workers and homemakers of tomorrow. If attention is directed to their physical, mental and moral welfare, the workers of tomorrow will be better fitted to enter the industrial world, and their contributions to industry will be that of greater efficiency." In sum, "the social welfare programs of TCI were, at least in part, designed to acclimate a rural people to a more urbanized environment and in the process, . . . create the type of stable, healthy—and non-unionized—work force desired by the company."[33]

In the Cahaba coal field, the medical arm of welfare capitalism assumed different forms. For example, TCI opened a hospital in Blocton in May 1910 to provide more timely health care to residents of north Bibb County. One Blocton newspaper reported that "when everything is in readiness it will be an up-to-date hospital with all the necessary equipments [*sic*], and patients that have heretofore had to be carried to Birmingham can be treated here with all the comfort that can be had in a larger city." Seven years later the hospital closed, but continued operation as a dispensary affiliated with the TCI Fairfield Hospital (later Lloyd Noland).[34]

At Piper, the LCCC established offices for a doctor and a dentist. The dentist was available only one or two days each week, but the physician, Dr. W. E. Stinson, made house calls and earned a reputation for keeping new mothers in bed for two weeks after giving birth. As Sewell reports: "Doctors swabbed sore throats with 'silver-nitrate,' set broken bones, [and] prescribed Calomel to be followed by Castor Oil or Epsom Salts . . . over-the-counter medicines included Black-Draught and Compound Cathartic pills." A "stoppage" of $1 per person per month financed medical care, and the same amount ensured hospital coverage, usually at Norwood or South Highlands.[35]

Similar arrangements characterized Margaret's health care program. The Alabama Fuel and Iron Company provided the doctors, and residents received hospital care in Birmingham. One veteran black miner experienced this medical care firsthand when he sustained an injury in the mines. When doctors determined that he would be unable to return to work, owner Charles F. DeBardeleben established a pension plan to provide for the miner and his wife. Commenting on the overall atmosphere at Margaret, the miner exclaimed: "There ain't no place like it this side of heaven; everyone here is happy."[36]

On the other hand, this miner's perspective belies the true intent of "Uncle Charlie's" benevolence. Perpetuating his father's legacy, Charles attempted to manipulate black racial pride to promote loyalty to the company. Previously, Henry F. DeBardeleben's concept of a "Negro Eden" promised black miners a "fair show" in the face of Jim Crow restrictions. In theory, black workers could maintain their own institutions and societies without interference from white miners. But, in practice, the DeBardelebens "corralled" their workers, subjecting them to company interference and control. Similarly, historian Brian Kelly depicts Charlie DeBardeleben as "the toughest" of the antiunion operators and describes Margaret as a "regular slave camp." Highlighting the coercive nature of welfare capitalism,

Kelly asserts that "coalfield paternalism bore the unmistakable signs of its lineage in antebellum agriculture."[37]

This aspect of social control is personified by the "mystery man" who frequented the mines at Acmar, Margaret, and Overton. Cloaked in a black robe streaked with white skeletonlike markings, this specter brandished claws and horns outlined with flashing lights. Designed to intimidate superstitious miners, the masquerade complemented the "organizer trap" designed by "Uncle Charlie." Dynamite and machine guns guarded access routes to the mines, and a 220-volt electrified fence surrounded DeBardeleben's house. In addition, billboards warned visitors to enter "at their own risk" and declared—"We Are 100% Non-Union And Proud Of It." A sign posted by the Acmar Welfare Society proclaimed:

NOTICE
We Are Americans
And Believe in American Principles
If You Are Here to Interfere
With Our Rights, This is
The Place to Turn Around.[38]

Another factor in maintaining a stable workforce involved construction of company housing. Operators elected to build and maintain houses due to the transient nature of the workforce. Also, mining families generally had neither capital nor inclination to build their own homes. Thus, company housing developed from necessity, but also served as an inducement for employees. Once companies completed construction of these houses, full occupancy became a priority. In turn, full usage of company structures usually translated into greater revenues derived from rent payments and company store sales.[39]

The Bureau of Labor Statistics revealed that only 15 percent of all company towns in the United States used the services of town-planning experts. In most cases, coal towns simply developed based on mining operations and the terrain. Nearly two-thirds of all Alabama miners lived in 3,617 company-owned houses. As Flynt indicates, these dwellings could range from one-room shacks to four-room houses, but the latter arrangement constituted the norm. Brandes suggests that the four-room design stems from British expectations of living space for an average family. In general, these houses contained two bedrooms, a kitchen, and a combination living and dining room. In some cases, depending on the size of the family, the living room could be converted into a third bedroom. At Blocton, company houses contained as many as five rooms, but Number 3 Town followed the typical one-story, four-room, wooden frame construction model. Similarly, the New Camp at Aldrich included houses with three to five rooms and tin roofs. The four-room standard existed at Margaret as well, and these structures rented for $6.90 per month (including water and electricity). At Piper, rents ranged from $5 to $10 (including utilities) due to the company's policy of increasing the rent by $1 with each change of tenancy.[40]

Original wooden-shingle roofs eventually gave way to composition shingles, and exterior walls consisted of roughhewn lumber in board and batten style. Few houses were painted at Blocton, but the superintendent's house constituted an exception. One resident commented: "I supposed it was the finest house in the land as I passed it when I was a boy." Operators at Aldrich, Margaret, and Piper used a variety of colors including blue, brown, gray, green, red, and white to paint workers' houses. Interior walls and ceilings were finished with smooth lumber, plaster, paint, or wallpaper. Coal-burning fireplaces provided limited heating during the winter months, and open doors and windows caught breezes during warmer periods. A tin bathtub usually stayed close to the kitchen stove during colder weather as well.[41]

Most company-owned houses included a sizeable yard, a fence, and at least some flowers and shrubs. Shade trees were scarce in Blocton, and Coleanor yards had no grass. Residents swept their dirt yards with brush brooms made from dogwood branches. Garden areas produced vegetables and fruits, and fences kept domesticated animals such as chickens, cows, pigs, and goats close by. These fences served a dual purpose in keeping roaming animals—grazing the open range—out of cultivated areas. The back yard provided a washing site complete with wash pot, tubs, rub board, and clothesline, and a short footpath led to the outdoor privy.[42]

Before the introduction of electricity in the 1920s, few houses benefited from running water. At Blocton, a gravity-fed system provided water from tanks on the hill above the coke ovens to a hydrant in the street. Piper residents drew their water from wells until electric power facilitated construction of a pumping station that supplied a stand pipe on each back porch. The pumping station transported water directly from the Cahaba River until state health laws required filtering. In 1928, the LCCC added a filtering plant near Piper Mine Number 2. The advent of electricity also introduced new appliances into coal towns, and refrigeration rendered traditional ice deliveries obsolete.[43]

Electricity was not the only catalyst for change in the coal towns of the 1920s. Religion and local churches also responded to new dynamics. With the departure of numerous European immigrants during World War I, Roman Catholic, Lutheran, and Jewish congregations experienced a decline in their overall population. At the same time, more evangelical denominations—Methodist, Baptist, Nazarene, Church of God, and various Pentecostal sects—gained greater acceptance. Even so, Shifflett notes that coal miners remained irregular church attendees. Church activities met spiritual and social needs in rural areas, but coal towns offered Sunday afternoon baseball, picnics, reunions, holiday festivals, and other events in competition with the church. Nevertheless, Shifflett argues that the church represented a "powerful instrument of social control." As a vital part of contentment sociology, organized religion attempted to reinforce the development of a satisfied, stable, and productive labor force. In his study of textile workers in Gastonia, North Carolina, Liston Pope asserts that the local churches reinforced public approval for the textile entrepreneurs and promoted their acceptance as the rightful leaders of their communities. In similar fashion, Shifflett contends that miners resisted attending church with "company men" and that, "feeling the church had abandoned them, [they] abandoned the 'company' churches."[44]

These models may hold true for the textile industry and for coal mining throughout Appalachia, but the prolific growth of churches in the Cahaba coal field suggests otherwise. Rather than supporting the idea that miners and their families resisted the church as a "company" institution, the opposite trend occurred in Cahaba. In effect, the plethora of congregations and the abundance of denominations reflect the racial and ethnic diversity of the Cahaba coal-mining communities.

By 1890, the Blocton area supported eight white and two black churches. Mt. Carmel, the oldest church in the Blocton area, began as a Missionary Baptist Church in 1869. Six years later, the Mt. Nebo Primitive Baptist Church organized in the sawmill community of Primitive Ridge. In 1883, the First Methodist Church of Tuscaloosa began a "mission" work, and Blocton residents met on the lower floor of the Odd Fellows Hall. This building, used for church, school, and social events, accommodated gamblers and drinkers on Saturday night and Methodist worshipers on Sunday morning. This Methodist congregation built the first church building in Blocton Number 2 Town in 1886 and then moved to Number 1 Town in 1900. By 1886, resident Catholics observed mass twice a month in private homes. Blocton's Italian community supported this mission of St. John Parish in Tuscaloosa, and they built the St. Francis of Assisi church building on Gunlock Hill at some point prior to 1900. The arrival of a resident priest in 1901 prompted construction of a rectory and a three-room school. The school, administered by the Benedictine Sisters from Sacred Heart Motherhouse in Cullman, enrolled 125 students. In 1887, Blocton residents established the Church of Our Savior Episcopal Mission and the Blocton Baptist Church. The Baptists built the first church building in Number 1 Town, but the Episcopalians never achieved more than "mission" status.[45]

Other white denominations consisted of Cumberland Presbyterians, Disciples of Christ, Independent Methodists, Welsh Congregationalists, Wesleyan Methodists, and Free Will Baptists. African American congregations included Peace Baptist, Hope Hill Baptist, Liberty Baptist, Friendship Baptist, and New Zion African Methodist Episcopal. Peace Baptist marks its beginnings in 1888, but

Adams notes that "the black community had been served by Baptist and Methodist churches built by the Cahaba Coal Company sometime in the mid-1880s." In addition, a sizeable Jewish population that comprised immigrants from Poland, Lithuania, and Russia, worshiped in private homes and local meeting halls in the mid-1890s and then built a synagogue in 1905.[46]

The residents of Piper, Coleanor, Aldrich, and Margaret were less diverse than those at Blocton. Piper and Coleanor organized three Methodist (two white, one black) and two Baptist (one white, one black) congregations to conduct worship, revivals, funerals, weddings, and other social functions. The white churches at Piper adopted "half-time preaching"—Methodists conducted services on two Sundays each month, and Baptists met on the other two. At Aldrich, black miners built the first church—Epsibeth—in 1873, about one-half mile from the mine entrance. White congregations of Baptists, Methodists, and Roman Catholics met on the second floor of the commissary building until William F. Aldrich constructed a church building in 1892. This community structure served as the school and social center in addition to providing space for several congregations of Methodists and Baptists. The white Aldrich Baptist Church was constituted in 1932, and the black Green Chapel Missionary Baptist Church was founded in 1936. At Margaret, St. Phillip Methodist Church and Beulah Baptist Church organized in 1917 and 1918, respectively. Similar to other towns, these congregations alternated services and shared a public building that doubled as the local school.[47]

Arguably more influential than the community churches, local schools often reinforced paternalistic policies. Brandes reports that, in many cases, companies constructed school buildings, hired teachers, paid salaries, and then made education as compulsory as practicable. Even when worker "stoppages" supplemented company funds, operators exercised "a certain control" over community education. In general, company executives believed that education held the key to a contented and stable workforce. Also, by educating children early in the ways of industry, company-sponsored schools prepared the next generation of labor. Consequently, businesses generally supported education through the eighth grade, but company-sponsored high schools remained scarce.[48]

As a rule, operators believed that efforts to educate the workforce should concentrate on children rather than adults. A basic assumption stated that "children were more pliable than adults, more 'educable.'" However, many companies established adult-education programs. For example, the Red Ash Coal Company collaborated with the Bibb County Board of Education to sponsor an "Opportunity School"—an adult education program initiated at the Number One Blocton School in 1929. In addition to increasing literacy among their workers, operators used classes to undercut union activity and to assimilate immigrant labor. Frequently, adult classes concentrated on "Americanization"—knowledge of the

English language, cleanliness, wearing American clothes, and eating American cooking.[49]

Another aspect of company-sponsored education focused on the local library. In addition to promoting increased literacy, operators encouraged their workers to read. A logical application of this concept found fruition in the employee magazine. Designed to create a cohesive organization and community, these company-sponsored publications espoused corporate policies as well as reporting births, marriages, vacations, promotions, and other items of interest to the employees. Brandes asserts that "contentment was a prime goal . . . translated into . . . a willingness to accept a business view of industrial conditions. . . . Another means of building contentment was by creating the impression that one belonged to what was really 'one big happy family.'"[50]

The TCI school at Blocton offers the best example of paternalistic education in the Cahaba coal field. This school, established shortly after TCI's purchase of the Blocton mines in 1892, provided classes for grades one through eleven. In spite of the lack of the traditional "senior year," the school established a reputation for quality that enabled students who completed the curriculum to gain admission to state colleges and universities. Remodeled in 1907 as a result of TCI's acquisition by U.S. Steel, the Blocton school retained its high standards for teaching and learning. Company funds maintained a teacher's home, or "teacherage"—a residence designed to provide "an example to the women of the community of how houses should be kept and run." The company also provided adequate supplies, financed indoor plumbing, and provided several extracurricular programs. In addition to the basic curriculum, boys learned techniques in gardening, and girls studied the "domestic sciences" such as cooking and sewing. The school supported teams in baseball, softball, field events, and speedball, and the company sponsored annual sports competitions in Birmingham.[51]

The TCI school did not constitute the only opportunity for education in Blocton, however. In 1890, R. W. Jones opened a school at the Odd Fellows Hall, and Miss Dovie Hayes taught at the Mt. Carmel Church. By 1897, Blocton supported four white and two black schools. White schools were located at Number 1 Town, Number 2 Town, Campbell Town, and Smith Hill. The more notable black school—the Cadle School—offered nine grades and was located on the Old Woodstock Road in "Number 3 Quarters." In general, these schools conducted six-month terms, and public money combined with individual fees (approximately 50 cents per pupil per month) paid teacher salaries.[52]

The West Blocton School opened in 1903 in a two-story, eight-room building. Enrolling 132 students in eight grades, this school received funding through the dispensary system. This arrangement earmarked profits from local liquor stores for roads and schools. Half of these revenues went to the county, but the other

Fig. 24. West Blocton School. (Photograph by John T. Weeks. Courtesy of Beverly Weeks Beasley.)

half could be used for local projects. The idea of linking education to liquor sales generated considerable debate, but a common argument stated that, "while the dispensary system is condemned by some, it is considered by many to be in the interest of good municipal government and appears to be the solution of the whiskey question for small towns." In spite of some opposition, the dispensary system apparently worked effectively until Bibb County converted to "dry" status in January 1909. West Blocton opened its high school in 1920, but with no twelfth grade. The community approved a school tax the following year, but sophomores and juniors paid a $45 tuition to fund their final or senior year. This policy continued until school officials added a twelfth grade in 1924.[53]

Piper and Coleanor maintained two separate schools for white students until they consolidated resources at Piper in 1912. The Piper-Coleanor School offered eight grades and conducted nine-month terms. High school students rode the L&N passenger train, the "Joe Lee," to and from high school at West Blocton. Black children attended grades one through nine on the east side of Piper, and older students attended high school in other communities. Piper added a white high school in 1930 and graduated ten classes from 1931 to 1940. Two sports fields—one for football and baseball, and another lighted field for basketball and tennis—supported the school athletic programs. Residents also used the school auditorium for plays, concerts, and other community gatherings; movies showed every Tuesday and Friday night.[54]

Similar arrangements developed with schools in Aldrich and Margaret. William F. Aldrich constructed two one-room, coal-heated buildings, one for whites and the other for black students. Segregated schools remained the norm at AF&IC schools located at Margaret, Lowgap, White's Chapel, and Acmar. Black students attended school in the local church building until completion of a separate school facility in the 1930s. The AF&IC furnished supplies for a nine-month term, and parent-teacher associations supplemented the formal curriculum with classes in gardening.[55]

In addition to the company store, applied medicine, housing, churches, and schools, leisure activities and recreation helped define the coal-town culture. For example, visiting—exchanging news and gossip—represented a favorite pastime. Often, waiting for mail delivery or for a train's arrival facilitated this activity. As Shifflett reveals, sometimes "just sitting and talking" provided its own form of recreation. Radio also emerged as a popular medium for entertainment. Adams reports that by the mid-1930s "almost every home [in Blocton] had a radio and a list of favorite programs." Other activities included picnicking, church socials, dances, quilting and crafts, apple-pickings, bean-stringings, hunting, fishing, and swimming. However, the most popular leisure activities centered around five general categories: sports, theater and shows, bands, fraternal societies and clubs, and events spawned by holiday celebrations or festive occasions.[56]

By far, baseball represented the most popular sport. Tennis, basketball, and football gained some support, but Shifflett identifies baseball as "the miners' sport." Miner recruitment and public relations affected both white and black company-sponsored teams. In many cases, operators created jobs for outstanding players, arranging for higher pay, better schedules, lighter work, and afternoons off. Coal miners worked hard, and baseball served as a diversion from the demanding and hazardous routine. Consequently, participants played with a passion that added intensity to the sport. Sunday afternoon games became community events complete with ice-cold lemonade and soft drinks. As the game's popularity increased, unofficial leagues organized games among various coal towns. For example, Blocton's team, a member of Birmingham's Southern League, played games against Aldrich, Bessemer, Montevallo, Randolph, and Selma. The Blocton nine also challenged the University of Alabama to a game in Tuscaloosa in 1892, thereby initiating a five-year series. At that time, approximately five hundred Blocton fans boarded the train to accompany their team to the Capstone.[57]

In September 1892, the Associated Press sponsored a round-by-round telegraphic report of the Sullivan-Corbett boxing match. "Spectators" bought tickets for 50 cents and gathered in Blocton's Odd Fellows Hall, intently following the fight through twenty-one rounds lasting one hour and twenty-three minutes. Although boxing attracted an enthusiastic audience, most of the theaters and meeting halls

Fig. 25. Baseball team in Aldrich, Alabama. (Courtesy of Marshall Goggins.)

served a more refined clientele. For example, the Blocton Philharmonic Society organized in 1893 even though local facilities proved inadequate. In 1907, the Theatorium opened in June, and the Opera House followed three months later. Both enterprises prospered, and the Opera House provided shows, moving pictures, plays, and vaudeville acts until 1925. The Theatorium sponsored Saturday night shows, plays by traveling companies, moving pictures, comedy acts, and musical concerts. In addition, the Pandora Theater, opened in 1913, offered nightly vaudeville shows with performing companies changing each Monday and Thursday night.[58]

Theatorium owner T. C. Tuggle rebuilt his facility after the great West Blocton fire of 1927, and the new West Blocton Theater provided "talking pictures" thereafter. This 675-seat theater had been configured for silent movies before 1927, and a player piano provided sound effects—train and steamboat whistles, sleigh bells, gongs, drums, and galloping horses—as needed. By 1930, manager Earl Philpot attracted children to his Saturday "egg matinees." For one fresh egg, a child gained admission to see a newsreel, comedy, serial, and full-length western. Philpot then sold the eggs to local grocery stores.[59]

The local theaters also supported various community bands. The Cahaba Coal Company first organized Cadle's Silver Cornet Band in 1891, providing uniforms

Fig. 26. Alabama Fuel and Iron Company Band in Margaret, Alabama. (Marie Butler, *Margaret, Alabama . . . and Now There's Gold: A Brief History of a Unique Community* [Birmingham: privately printed, 1989], 28.)

for all nineteen members. The company also sponsored a band excursion to New Orleans that ultimately included approximately three hundred Blocton residents. After the TCI acquisition, the Blocton Number 2 TCI Band replaced Cadle's musicians, and this group eventually converted into the "Little Italy Band." The Suttletown String Band originated in 1892, and the Blocton Home Band organized in 1913. The latter group performed its first concert in the Opera House and participated regularly in town celebrations, dances, and parades. Unfortunately, these musicians lost company sponsorship as a result of strikes from 1919 to 1921. Nevertheless, the West Blocton Band remained active until 1923, and the West Blocton String Band performed three times per week on Birmingham's WAPI radio station in 1929.[60]

Blocton's fraternal orders originated around 1889. By 1892, four organizations collaborated with TCI to construct the Park of the Four Orders, or TCI Park. This project, sponsored by the Order of Odd Fellows, the Masonic Lodge, the Order of Red Men, and the Knights of Pythias, provided a gathering place for community patriotic celebrations, ball games, political rallies, band concerts, circuses, and carnivals. African American residents participated in similar orders, namely the Masonic Philanthropic Lodge, the Order of Odd Fellows, the Knights of Pythias, and the Order of the Eastern Star. Other community organizations included the

United Confederate Veterans, the United Daughters of the Confederacy, the Ladies Needle and Social Club, the Three Arts Club, the Book Club, and a debate society known as the Prospectors Club.[61]

Finally, community events provided various forms of recreation and entertainment. The circus came to Blocton first in 1896, and various companies returned with elephants, monkeys, and clowns to entertain residents with parades and performances at Four Orders Park. Annual events included an "Old Soldiers" gathering sponsored by the Confederate Veterans, a Fiddlers Convention, and homecoming. Activities included picnics, baseball games, political speeches, live music, and dancing. At Margaret, similar events included a barbecue, games, a horseshoe tournament, and a greased pig contest. At Piper, weekend festivities and Fourth of July celebrations constituted all-day affairs with baseball, barbecue, ice cream, and dancing. Also at Piper and Coleanor, a Christmas tree program at the Methodist church reflected a community effort. Each child received a gift and a bag of fruit, and the ringing of the church bell followed the Christmas pageant. Similarly, Margaret residents gathered at the annual Christmas program, and owner Charles DeBardeleben provided for needy families as Santa Claus distributed toys and fruit baskets to the children.[62]

Thus, welfare capitalism created an environment that offered many opportunities for coal-town residents. Enduring hard work and dangerous circumstances, miners and their families warmed to the benevolent offerings of their employers. Whether exchanging greetings at the commissary, responding to the doctor's care, visiting among neighboring houses, worshiping in the local church, learning in the community school, or celebrating at the holiday festival, they benefited from their surroundings. On the other hand, everything ultimately belonged to the company and fell within the operators' control as company cessation of support for the Blocton Home Band demonstrated.

Aspects of this benevolent environment prompt historian David Brody to conclude that, if not for the Great Depression, "the affection of workers for welfare capitalism [in northern manufacturing corporations] would have continued unabated." On the contrary, the nature of this two-edged sword causes Brandes to argue that the New Deal offered American workers a choice, and "they shunned welfare capitalism in favor of unionism." He contends that when operators approached their employees from a fatherly perspective, they relegated those workers to an inferior, childlike role. By providing services such as commissary goods, medicine, housing, church and school buildings, and various forms of recreation, officials assumed that laborers could not provide for themselves. In turn, when employees accepted such benevolence, they acknowledged and endorsed their subordinate position. In sum, paternalism stripped workers of their self-respect while ensuring social control for the operators.[63]

Examples of benevolent paternalism at Aldrich, Piper, and Margaret must be viewed as aberrations. Relative calm and contentment in the early twentieth century belied the constant struggle between labor and management. In fact, the "peaceful" environment of the paternalistic period resulted from devastating defeats for organized labor at the hands of antiunion forces. Failed attempts to organize miners and promote collective action—in 1890, 1894, 1904, 1908, and 1920–1921—left coal workers at the mercy of company management. Miners most discontented with such paternalism probably left or were forced out during these strikes. In essence, welfare capitalism emerged as a method for social control. Humanitarian gestures and benevolent policies merely reflected the operators' total domination of the labor force after 1908.

7
UNIONISM
1878–1935

During the widespread coal strike of 1920–1921, miners at Piper received eviction notices, and the United Mine Workers of America (UMWA) erected a tent city about five miles to the east. The union provided some food, but the poor quality and meager quantities prompted family members to name the encampment "Esau" in reference to that biblical character's anguish over his lost birthright. Denied access to the company store and short on cash, many miners patronized a country store at Hawkinsville, about three miles away. The proprietor, E. C. Day, knew most of the miners and extended credit with the understanding that the union would cover the charges. Ultimately, only two families repaid their debts. When the union lost the strike and the miners moved away, Day was forced to close his store. Thereafter, he worked for his former competitor, the Little Cahaba Coal Company commissary.[1]

Such action by government-supported coal operators constituted the norm throughout Alabama's coal district until unionism came to stay in the mid-1930s. President Franklin D. Roosevelt's New Deal legislation infused legitimacy into the hard-fought labor movement. Specifically, Section 7a of the National Industrial Recovery Act, signed into law in June 1933, authorized American workers "to organize and bargain collectively through representatives of their own choosing." Two years later, the U.S. Supreme Court declared the National Industrial Recovery Act unconstitutional, but Congress countered that decision by adopting the National Labor Relations Act or Wagner Act. The revised legislation incorporated many provisions from Section 7a and "created a strong enforcing body in the National Labor Relations Board."[2]

The advent of union recognition and collective bargaining marked the end of the paternalistic period. Stuart Brandes attributes the demise of welfare capitalism to an "aggressively antagonistic government." He contends that the New Deal denoted opposition to various aspects of welfarism and that the National Recovery Administration promoted tenets of fair competition that undermined paternalistic practices. In addition, Rikard argues that "'avoidance of trade unionism' was the 'central purpose of welfare capitalism.'" Therefore, promotion of one strategy constituted decline for the other. Reversing a seventy-year, pro-business trend, collective bargaining revealed the ultimate failure of welfare capitalism. As sociologist Richard Edwards observes, paternalistic policies sought to control labor by offering incentives and rewards to loyal employees. However, continual militancy among workers "indicated that welfarism did not do the job." Rather, as company influence declined, operators reduced or eliminated some programs while they attempted to shift responsibility for other benefits to various government agencies.[3]

Even with federal sanction, organization efforts faced considerable opposition from local operators. For example, at Piper and Coleanor approximately two-thirds of resident miners joined the UMWA. Still, officials of the Little Cahaba Coal Company refused to recognize union representatives or to negotiate a revised contract. Consequently, three hundred miners (two hundred at Piper-Coleanor, one hundred from the Bessemer Coal, Iron, and Land Company at Belle Ellen) failed to report for work on February 20, 1934. Showing support for their fellow workers, other miners at Piper and Coleanor walked off the job. Striking miners demanded a new contract that included a check-off system that provided for automatic payroll deduction of union dues. Picket lines formed at mine entrances and at the Cahaba River bridge in an attempt to prevent strikebreakers from disrupting the protest.

Three days later, former Tarrant City police chief A. M. (Mike) Self and fifteen "heavily-armed" guards arrived by bus at Coleanor. Ostensibly, the Little Cahaba Coal Company hired Self and his associates to "protect company property, keep peace, [and] preserve order." However, Self's reputation for employing harsh and extreme strikebreaking methods prompted miners to prepare countermeasures. Arming themselves with various types of firearms and clubs, miners surrounded the company office and the local boarding house while Self and his men ate supper. After entering the boarding house, miners disarmed the paramilitary group, led them outside to the waiting bus, and then escorted the vehicle out of town.[4]

As the strike spread to Belle Ellen, Boothton, and Straven, Governor Benjamin M. Miller (1931–1935) deployed Alabama National Guard units to the Cahaba field in an effort to maintain order. With machine guns protruding from train car windows, National Guard troops encountered miners armed with shotguns and sundry weapons upon their arrival at Piper. Attempting to defuse the volatile situation, the detachment commander, Major Benjamin Smith, voiced his desire to keep the peace. Instructed to maintain order, Smith adopted a sense of objectivity and refused to protect nonunion or "scab" labor. Consequently, the mines remained closed, and negotiations ensued. Payroll "stoppages" emerged as the primary point of contention as miners complained that the company liberally deducted everything but union dues from their paychecks. Therefore, the check-off system constituted a relatively simple measure supported by precedents. This

1934 strike marked the first time that federal and state authorities offered a supportive environment for collective bargaining. Consequently, after three weeks of negotiations, operators and miners reached a settlement. Adopted on March 16, the new agreement instituted semimonthly paydays (in cash or check), an eight-hour work day and a forty-hour workweek, and voluntary check-off for union dues. Thus, Cahaba miners experienced their first success in the art of collective bargaining.[5]

In contrast, employees of Charles F. DeBardeleben's Alabama Fuel and Iron Company resisted unionization. Generally, union representatives depicted DeBardeleben as a cruel dictator who used armed guards to subjugate miners, maintain order, and deprive workers of happiness. On the other hand, most Alabama Fuel and Iron Company miners remained outside the union and pledged their full support to the company. In fact, "Uncle Charlie" paid his workers on par with union wages, and a local billboard discouraged efforts by union organizers. These conflicting issues came to a head in the fall of 1935 when armed UMWA miners advanced on Margaret and other nearby mining towns in an attempt to coerce union membership. Loyal residents emplaced machine guns near the mines, and a firefight ensued. One UMWA sympathizer died, and several others were wounded in the melee. Ultimately acquitted of criminal charges, DeBardeleben initiated a countersuit that asserted the right of a company and its employees to defend their homes and jobs. Addressing the local welfare societies, "Uncle Charlie" vowed to "close his mines and let the grass grow up around them before he would resort to operating them as union properties."[6]

The contrasting responses to unionism manifested at Piper-Coleanor and at Margaret typified the continual struggle enmeshed in seventy years of mining in the Cahaba region. Sociologist Shaunna L. Scott attributes such conflict to different points of view. In her study of coal miners in Harlan County, Kentucky, she discovered that workers generally viewed company success as the means to an end. In effect, profits allowed companies to achieve the operators' ultimate goal—caring for their workers and communities. From the other perspective, Scott observes that coal companies exist for only one reason—to make money. Consequently, from the operators' viewpoint, profits represent an end in themselves, and jobs constitute a mere "by-product of the original intent."[7]

Tracing its roots to the Greenback-Labor Party (GLP) and to the Knights of Labor, organized labor experienced numerous setbacks and disappointments in its quest for recognition. As Crandall Shifflett notes, miners who hand-loaded coal during the last quarter of the nineteenth century were conditioned by rugged, rural circumstances to endure a relatively high degree of hardship. Coming from harsh, primitive environments, "they expected to work long hours, to perform hard physical labor, [and] to live in modest frame houses. . . . Most were happy just to have a steady well-paying job." In effect, workers' low expectations opened the way for ambitious operators to exploit the labor force in an attempt to maximize profits.[8]

Because of the independence and diversity inherent in the coal-mining population, the GLP sought a common bond that would generate a sense of class consciousness. With relatively equal numbers of white and black miners, racial issues bore preeminent importance within southern society. As labor historian Daniel Letwin observes, "The color line bisected virtually all areas of the miners' lives." Within this context, operators frequently promoted racial tension to subvert unification efforts. However, the coal-mining industry exhibited a degree of compatibility and acceptance among white and black workers. Sharing the risks and hardships of a demanding and dangerous occupation, Alabama's coal miners adhered to the concept of "equal pay for equal work." As a result, labor organizing efforts throughout the coal fields promoted interracial cooperation, and "the labor movement in Alabama . . . centered among the coal miners."[9]

Initiating recruitment efforts in 1878, Greenbackism complemented attempts to organize workers by combining union rhetoric with evangelical zeal. Whites and blacks joined the Greenback movement, but letters written to the *National Labor Tribune* in Pittsburgh indicate that the GLP in Alabama gained most of its support from black miners. Historian Herbert Gutman suggests that post-Reconstruction discontent prompted many blacks to abandon the Republican Party in search of other means of reform. Correspondence from the Helena mines in July 1878 reported a significant increase in membership within the National Greenback Party.[10]

Much of this popularity and growth may be attributed to the recruiting efforts of Willis Johnson Thomas, a twenty-nine-year-old black miner headquartered at the Jefferson Mines. Speaking at various Greenback-Labor clubs throughout the Birmingham District, Thomas personified the organizational movement. One observer commented that "he took the [B]ible and made some quotations. He spoke of Moses, of Joseph and others. He spoke of Christ, and his love for the people and indeed the meeting looked, for a moment, more like a . . . camp-meeting than a political gathering." Previously segregated clubs combined their efforts, conducted interracial meetings, and elected biracial committees. In general, four Greenback officials—two black and two white—negotiated contracts, supervised strikes, and settled work stoppages. The following year, Michael Moran preached the Greenback "gospel" at the Helena Mines, and he issued the following challenge to all workers: "Fellow miners and workingmen, the time has come for us to act together. Our fellow toilers are moving in one grand mass all over this noble land of ours (it ought to be ours, only for our cowardice). Therefore, I ask you, men of Alabama, to be up and doing your duty. The time is coming when labor will assert its rights all over the land, and I hope all over the earth, for 'the earth is the Lord's and the fullness thereof,' but it seems our paternal government don't [*sic*] think so."[11]

In spite of its widespread appeal among Alabama's coal miners, the Greenback-Labor movement was short-lived. By 1882, the GLP succumbed to the increasing popularity of the Noble and Holy Order of the Knights of Labor. The Knights, organized in Alabama in 1879, gained a foothold in the Cahaba coal field at Helena. Many former members of the GLP, such as Thomas and Moran, continued their organizing efforts under the auspices of the Knights of Labor. Establishing local assemblies at Aldrich, Blocton, and Henry Ellen, the Knights enjoyed success throughout the Cahaba field as well as in the Birmingham District. Building social ties within mining communities, leaders organized parades and holiday celebrations in addition to conducting funerals and publishing death condolences. Mobilizing the organization politically, local officials created mine committees, endorsed mine inspector candidates, monitored working conditions and miner populations, supported cooperative businesses, and conducted strikes. As one miner put it, the Knights of Labor existed "to resist the tyranny imposed . . . by the selfish and greedy, who think that the employer ought to have it all and the poor, hard-working miners nothing."[12]

Building on the GLP experience, the Knights of Labor focused on "education, arbitration, and economic cooperation." More specifically, the organization advocated an eight-hour day, abolition of contract and convict labor, arbitration of disputes, monetary redistribution, and land reform. Moreover, the Knights encouraged miners to seek self-education and self-improvement. Discouraging strikes, leaders promoted cooperation and mutual support between labor and management. The Knights did support a number of independent strikes throughout the decade, but ultimately failed to provide sufficient centralized control. The Knights of Labor peaked in popularity in 1886, but then experienced a decline that coincided with a nationwide trend.[13]

From 1887 to 1890, the Knights collaborated with other organizations in an attempt to better support miners' interests. Affiliations with the Alabama Union Labor Party (later the Labor Party of Alabama), the National Federation of Miners and Mine Laborers, and the Miners' Trade Council (formerly the Miners' and Mine Laborers' State Trades Council of Alabama) failed to rejuvenate the Knights' influence. Locally, assemblies suffered from unsuccessful strikes and conflicts between Knights leaders and discontented miners. Facing statewide resistance to unionization, opposition from Democratic politicians and New South boosters, and distractions based on racial issues, the Knights of Labor foundered in Alabama.[14]

Thus, initial attempts to organize miners in central Alabama failed due to inadequate centralized leadership. Numerous strikes spanned the 1880s, but each one suffered from geographical and political isolation. Still, the GLP and the Knights of Labor codified many of the miners' grievances and provided a foundation for subsequent labor movements. For example, miners complained that they received no compensation for "dead work"—tasks such as shooting rock, laying tracks, constructing guardrails, timbering roofs, clearing cave-ins, driving air courses, and hauling coal to tram cars. In addition, time spent in transport between the mine opening and the coal face generated no income. From the miners' standpoint, each of these tasks contributed to the overall production of coal.

On the contrary, operators contended that coal alone generated profits, therefore miners needed to work as efficiently as possible to maximize output. Because miners received payment only for coal tonnage, "dead work" constituted a moot issue for operators. In some cases, operators paid workers for a 2,240-pound ton, then sold the coal based on a 2,000-pound ton. Miners accused their employers of other transgressions: using faulty scales, increasing the size of coal trams, screening coal before weighing, and deliberately refusing to give proper credit for a miner's work.

Other grievances centered around the concept of a "miner's freedom." As a rule, coal miners exercised a degree of autonomy. Once supervisors assigned rooms and trams, each miner operated independently to extract his portion of coal. Because of this egalitarianism, miners developed the cardinal rule of "equal pay for equal work." Irregular work schedules allowed miners to supplement their mining income with farming or other employment, and restless miners could move from one mine to another within the same general area. In many cases, miners could be absent for up to three days without fear of retribution.

Even though exercising a degree of freedom, miners also experienced an element of helplessness and exploitation. Working conditions were atrocious, and the lack of safety procedures resulted in frequent accidents or explosions that killed some miners and left others injured or maimed. Continually trying to minimize operating costs, company officials refused to spend necessary funds to improve conditions within the mines. Furthermore, operators often determined wages based on a sliding scale. In essence, as the selling price of pig iron fluctuated, miners' wages reflected the rise and fall of the market.

From the miners' perspective, the most notorious elements of exploitation involved the use of convict labor and the subcontracting system. Obviously, convicts provided a ready-made source of strikebreakers, and operators used both state and county prisoners for that purpose. Miners had formed the Anti-Convict League in 1885, but efforts to curb the use of convict labor proved ineffective. The subcontracting system also served to undermine miner solidarity and reduced production costs. In this arrangement, operators contracted with a miner (usually white) who would hire up to ten laborers (usually black). Then, under the subcontractor's supervision, each laborer would dig his designated amount of coal. Receiving $3 to $4 per day for each laborer's output, the subcontractor paid each worker $1 and pocketed the rest. Free miners opposed this practice because it violated the egalitarian rule, promoted exploitation among miners, produced a surplus of unskilled workers, and generated preferential treatment.[15]

Thus, both the Greenbackers and the Knights of Labor contributed to an interracial movement that combined labor organization and political activism. Within this framework, the UMWA emerged as the predominant labor organization of the 1890s.

Organized in January 1890, in Columbus, Ohio, the UMWA committed itself to "educating all mine workers in America to realize the necessity of unity of action and purpose, in demanding and securing by lawful means the just fruits of our toil." Configured as an industrial union, the UMWA sought to organize miners as well as other mine workers. Because most black miners in Alabama worked as subcontractors or laborers, union organizers adopted the interracial stance of their predecessors. In fact, official UMWA policy stipulated that "no local union or assembly is justified in discriminating against any person in securing or retaining work, because of their African descent." Consequently, black miners occupied numerous positions such as vice president, executive board member, grievance committee member, checkweighman, and district or national convention delegate. In many cases, integrated local organizations reserved a number of leadership posts for black members. Even so, interracialism did not prove absolute. Rather, in spite of common class interests, white and black miners adhered to the social norms of segregation. One observer remarked that "while white and colored miners worked in the same mines, and maybe even in adjoining rooms they will not ride even on a work-train with their dirty mining clothes on together."[16]

By mid-1890, local UMWA organizations existed in Bibb, Jefferson, and Shelby Counties. Enjoying success in its initial organizational drive and in the adoption of Alabama's first Mine Inspection Act, the UMWA invited operators to discuss a revised national wage scale. However, most operators refused to recognize the union as a legitimate negotiating entity, and only a few independent owners participated in the parley. Consequently, UMWA officials called a strike to begin on November 29. Maximizing the effect by shutting down mines as winter demands increased, more than 3,000 miners initiated the walkout. Supporting demands for an increased wage, approximately 15,000 more joined the strike by year's end.[17]

In response, operators hired armed guards who immediately evicted miners and their families from company housing. Moreover, Henry F. DeBardeleben introduced a new tactic to the coal fields. Hiring numerous black workers as strikebreakers, he fired all miners affiliated with the UMWA and reopened the mines with his new labor force. The UMWA countered with a $500 donation from the national convention, but that token amount proved ineffective in relieving the miners' plight. As a result, the strike ended in mid-January 1891, and the UMWA vanished from the Alabama coal fields. As historians David Ward and William Rogers report, "the Alabama UMW moved from substance to shadow and into oblivion. It disappeared almost as completely as if it had never been organized in the state."[18]

Although brief, the strike of 1890 set the tone for labor conflicts for the next four decades. Common class interests brought coal miners together in interracial labor organizations, but operators used black strikebreakers to force a racial wedge between white and black workers. Unity persisted among the miners, but they were unprepared to deal with evictions, armed guards, and negative press coverage. Coal prices increased during the next two years, however, and general prosperity generated little need for collective action.[19]

The Panic of 1893 significantly altered that state of affairs. As coal production declined by 72 percent, numerous miners faced wage reductions and unemployment. On the other hand, operators relied on convict labor to maintain steady output at minimal production costs. As a result, miner representatives met in Bessemer in October to organize the United Mine Workers of Alabama (UMW-Alabama). Although similar to the UMWA in configuration and function, this state-based union remained separate from the national organization. Apparently, Alabama miners believed that they could affect their own destiny more easily through the smaller institution. But hard times persisted, and many miners, facing financial straits, either left the state for other mining areas or refused to affiliate with the local union.[20]

Those miners who remained in Alabama faced a five-cent-per-ton wage cut by mid-1893. Successful negotiations, though, mitigated the impact of that reduction, and many miners signed new contracts, valid through June 30, 1894. Meanwhile, operators sought to reduce expenses, and monetary shortages forced many companies to pay their miners in scrip. Miners could receive full value at the company store, but conversion to cash resulted in a 25 percent discount. This method of payment may have appeared practical to the operators, but many miners accused company officials of financial exploitation.

When the UMW-Alabama opened its convention in Birmingham in March 1894, officials advocated a spirit of cooperation between labor and management. Developing a proposal that included "reasonable concessions," union delegates agreed to a 10 percent wage cut effective December 31, 1894. These reductions in pay were contingent on reciprocal actions on the part of the operators. The UMW-Alabama proposed that companies weigh all coal before dumping, authorize miners to appoint their own checkweighman, and reduce house rents, store prices, and medical costs. The operators' counterproposal called for a 22.5 percent wage cut along with various cost reductions not exceeding 10 percent. Deeming the counter offer unacceptable, the UMW-Alabama voted to strike.[21]

Approximately six thousand Alabama coal miners struck on April 14, 1894. Encouraged by a nationwide strike initiated by the UMWA a week later, another two thousand miners joined the protest in Alabama. Letwin asserts that "although the Alabama strikers were not formally linked with the national strike, the latter gave their efforts a heightened sense of purpose." The company response proved typical; operators used convict labor and black strikebreakers ("scabs") to continue coal production.[22]

As tension mounted in the strike region, miners frequently resorted to violence in an attempt to prevent scabs from crossing picket lines. After a meeting with Governor Thomas G. Jones, the UMW-Alabama executive committee denounced "any act of violence that has been committed by any party or parties." Mining officials desired "to win a bloodless victory, and any acts of intimidation or violence will necessarily be detrimental to our cause." Nevertheless, miners continued to take matters into their own hands. Unable to stop scab labor, saboteurs resorted to dynamiting or burning railroad trestles. Destroying approximately one per day, these vigilantes sought to interdict coal shipments by interrupting transportation links. Ultimately, Governor Jones dispatched National Guardsmen and Pinkerton detectives to the strike zone to maintain order. UMWA historian Maier B. Fox contends that Jones "sent in troops for protection when there was no violence" and that he placed "responsibility for the maintenance of order on the miners." In contrast to Jones's perception, reports filed by the Pinkertons indicated that "the miners were peaceful but committed to the success of the strike."[23]

The national strike ended in mid-June, but Alabama miners continued their protest. In an attempt to undermine labor solidarity, DeBardeleben, now vice president of the Tennessee Coal, Iron, and Railroad Company (TCI), advertised free transportation to northern coal fields. Approximately one hundred miners accepted his offer that same day, but union officials countered with an innovation of their own. Courting support from Eugene V. Debs's American Railroad Union, the UMW-Alabama encouraged a railroad worker strike as a legitimate way of interdicting transportation routes. After a concerted membership drive throughout the Birmingham area, the American Railroad Union struck on July 7. Governor Jones intervened immediately, however, and the railroad strike lasted only four days.

Ostensibly, the strike ended at that juncture, and National Guard forces left the Birmingham District on July 16. But, more violence erupted almost immediately, and some mining communities continued their protests. Demonstrations persisted at Blocton where relative calm had prevailed. In early July, Governor Jones deployed a National Guard detachment to that community, and TCI officials increased pressure on striking miners. Directed to evacuate their houses within ten days, Blocton miners arranged a meeting with the governor himself. After listening to their explanations, Jones informed them that he considered the strike hopeless and that "its continuance meant lawlessness." He pledged to "protect other laborers" and implied that the state government would support TCI's efforts to reopen the mines with nonunion workers. He also announced that National Guard units would remain in the area until all strike activity ceased.[24]

Miners at Blocton and throughout central Alabama held one last thread of hope. Gubernatorial candidate Reuben Kolb supported the miners and denounced the state's use of convicts, black strikebreakers, and National Guard units. Jones's handpicked successor, William C. Oates, defended the governor's actions, citing outbreaks of violence as justification for the use of force. Oates won the August 8 election, thereby dashing the miners' hopes for a favorable settlement. Having persisted in their protest for four long months, miners accepted company terms and returned to work.[25]

Governor Jones described the strike of 1894 as the "most formidable and threatening commotion in the history of this State in times of peace." Organizing the largest labor demonstration in the nineteenth century, the UMW-Alabama survived four months of protest against overwhelming odds. Miners faced "inexhaustible reserves of convicts and strikebreakers, a hostile governor, a depressed economy, and an ambivalent public." Still, they failed to curtail production, obtain an acceptable wage, or gain union recognition. Furthermore, and perhaps more subtly, the 1894 strike changed the demographics of Alabama's coal-mining industry. As many white miners deserted the central Alabama fields, many of the imported strikebreakers remained. Consequently, the percentage of black miners increased, and interracial cooperation became more critical to collective action.[26]

Letwin observes that "unionism neither blossomed nor vanished during the mid-1890s." In fact, a lackluster economy and memories of the 1894 failure provided little motivation for organizing labor. Finally, an economic upswing in 1897 improved the outlook for union organizers. Success in midwestern coal fields encouraged the UMWA to renew its recruiting efforts in Alabama, and the UMW-Alabama transformed into its previously designated role as District 20 within the UMWA. Stimulated by the Spanish-American War, demand increased in 1898, and a degree of prosperity returned to the coal fields. As a result, operators worked openly with UMWA representatives. From 1897 to 1903, the union experienced virtual recognition as it negotiated annual contracts. Furthermore, the UMWA constituted the core element in the newly formed Alabama Federation of Labor. Established in 1900, this organization "sought to unite unions and central labor councils throughout the state in one powerful organization which could present a united front in labor's demands for favorable political action."[27]

Adoption of the 1901 Constitution implemented Jim Crow policies, however, and an economic decline in 1902 terminated the six-year period of goodwill. Contract negotiations failed in 1903 when operators used growing popular support for Jim Crow legislation as leverage in opposing the interracial UMWA. In May 1904, the union initiated a strike, but operators had stockpiled coal and coke in anticipation of an impasse. Resorting to past techniques, operators evicted miners from company housing and imported scab labor consisting of black workers and some European immigrants. Moreover, federal judge (and former governor) Thomas G. Jones supported the operators by issuing numerous court orders and injunctions. As local courts upheld company eviction policies, many miners were arrested for trespassing on company property. Meanwhile, the UMWA infused the strike zone with more than $500,000 in an attempt to provide basic services for its striking miners. "Tent cities" offered food and shelter for evicted miners and their families, and the union organized recreational activities such as ballgames, dances, parades, music, and speeches.[28]

Even though the strike lasted until August 1906, tempers flared and violence ensued within six months. For example, conflict in Blocton resulted in the murder of Superintendent W. S. Lang in November. Forsaking the moral high ground, miners eventually succumbed once more to operator pressures. Rikard states that "officially the strike lasted until August 1906, but effectively it ended by the fall of 1904." She asserts further that the miners' defeat in the 1904 strike marked "the first step in eliminating the UMW[A] from the Alabama coal industry."[29]

In fact, District 20 membership hovered between 3,500 and 4,000 from 1905 to 1907. Union activities remained relatively small-scale as the UMWA primarily represented miners employed by independent commercial operators. Coincidentally, Jim Crowism became more firmly entrenched, and many companies initiated elements of welfare capitalism in their coal-mining communities. Basking in the victory of 1904, operators ignored union contracts, reduced wages, and fired workers affiliated with the union. When company officials threatened to reduce wages by 10 percent, the UMWA called a strike for July 6, 1908.[30]

Facing possible extinction in Alabama, the UMWA launched a concerted effort to reorganize and mobilize the coal fields. District 20 president J. R. Kennamer identified the impending struggle as a "life or death" situation. Adding eight thousand new members in two weeks, the Alabama branch of the UMWA swelled to twelve thousand. Ironically, many former strikebreakers—black miners and European immigrants—joined the ranks of the union. Resorting to familiar tactics, operators relied on convict miners, imported scab labor, and petitioned the governor for state troops. Governor B. B. Comer ordered one thousand National Guardsmen to central Alabama to restore order. But, unlike previous strikes, these guardsmen developed an affinity for the miners and shared their austere existence.[31]

Frustrated by the influx of strikebreakers, some miners at Blocton resorted to violence. On Sunday, August 9, a small band of miners removed crossties from the Louisville & Nashville Railroad line in an attempt to prevent the arrival of non-union workers. Firing into the last two passenger cars, the Blocton miners killed three laborers and wounded several others. Officials arrested twenty-seven black and Slavic miners, three of whom were convicted of murder and sentenced to life imprisonment.[32]

Meanwhile, TCI officials evicted all striking miners, and the UMWA erected tents on leased farmland. Even though these encampments designated segregated living areas, local residents took offense to interaction among whites and blacks. Focusing on the UMWA's interracial tradition, operators, politicians, and newspaper editors accused the union of promoting "social equality." In addition, the newly formed Alabama Coal Operators' Association (ACOA) spoke of "bands of armed strikers, black and white, roving the district." Governor Comer responded by prohibiting public meetings. Citing unsanitary conditions as the motivating force, he also ordered National Guardsmen to cut down the tents and to prohibit construction of more "tented camps by the coal miners." At the same time, he summoned union leaders to a conference and issued an ultimatum. Threatening to convene a special session of the legislature to modify vagrancy laws, Comer instructed the miners to call off the strike. Implying that every striking miner could be arrested and then employed under the convict-lease system, the governor ordered the miners back to work. Buckling once more in the face of overwhelming odds, the UMWA terminated the strike on August 31. Having spent more than $400,000 in two months, the union faced total defeat and ceased to exist in the Alabama coal fields. As Fox states, "By the end of August, the UMWA was little more than a memory in Alabama." Brian Kelly agrees: "The strike's outcome had fundamentally altered the balance of power and settled for the foreseeable future the question of who would rule the coalfields."[33]

Miners who remained in the Alabama coal fields after the devastating loss of 1908 dealt with issues consistent with their nineteenth-century counterparts. Their list of grievances included "low pay and myriad deductions, poor ventilation and frequent mine explosions, cheating on weights and relentless dockage, overrecruitment and underemployment, decrepit housing and coercive commissaries, arbitrary mine bosses and repression of unionism." Moreover, union leaders continued to fight the public relations battle by insisting that labor's struggle consisted of industrial concerns rather than social (i.e., racial) issues.[34]

In contrast, operators continued to accuse the UMWA of advocating social equality for black miners. In addition to denying union recognition, company officials maintained control over the labor force by adopting aspects of welfare capitalism, employing a private police force, and maintaining a "blacklist" of all union supporters. New work rules increased pay deductions, pressured workers to patronize the company store, restricted local movement, and appointed company-approved checkweighmen. In addition to firing anyone with ties to the union, many companies hired private detectives to monitor the conversations and activities of their workers. Operators exerted their dominance in other ways as well. For example, at Margaret, a "shackrouster" or "company devil" checked on absent miners and coerced them into working if practicable. On a larger scale, Henry Bedham and W. S. Lovell, both ACOA members, purchased the Montevallo Coal Mining Company in 1912. At that time, the operation at Aldrich constituted the only remnant of UMWA influence in central Alabama. When the owners proposed changes to the current contract, union miners struck. The operators immediately replaced the union miners with convicts, and the Aldrich mines became nonunion.[35]

Even though exercising total control after 1908, operators may have realized that their victory constituted a pyrrhic one. On one hand, annual reports celebrated the lack of union activity; on the other hand, companies experienced a perennial problem of labor supply. Many skilled miners (mostly white) opted to leave Alabama for northern and midwestern coal fields, and unskilled strikebreakers increased production costs due to inefficiency. The search for skilled workers focused initially on European immigrants, but noncompetitive wage scales and a segregated social order deterred widespread migration southward. Therefore, operators sought cheap, black labor to reduce costs and to minimize the threat of unionism.[36]

Unfortunately for the companies, black miners began to leave the coal fields as well. The onset of World War I in Europe initiated a chain reaction that further siphoned skilled labor from central Alabama mines. When European immigrants returned to their native lands, northern industrialists experienced a labor shortage. Consequently, they looked to southern industry to recruit a new labor force. Seeking better pay and working conditions and eager to leave Jim Crowism behind, southern black workers moved to northern coal fields. The Great Migration depleted the black mining population and forced operators to hire and train unskilled laborers from the Black Belt. Many African American ministers and middle-class leaders, often with economic ties to coal operators, labeled the UMWA as the "white man's union" and encouraged black workers to "stay put." Nevertheless, letters from family members and friends encouraged the move northward, and significant numbers of workers—white and black—left the South.[37]

After a decade of subordination and outward migration, coal miners experienced a resurgence of autonomy within the Alabama coal fields. As markets expanded due to the World War, miners exercised their newfound leverage to negotiate better wages, working conditions, and overall treatment. At the same time, operators continued to use the blacklist to discourage union activity and to maintain an open shop. The UMWA responded with a recruitment drive in the spring and early summer of 1917 that resulted in 128 new locals. Spreading across the Birmingham District and the Cahaba coal field with "revival camp enthusiasm," the reorganization effort claimed twenty-three thousand union members among the district's twenty-five thousand miners.[38]

Renewed union activity coincided with American entry into the war. Concerns for national defense translated into more direct involvement in labor affairs by the federal government. Emboldened by the Wilson administration's commitment to peaceful labor relations, the UMWA issued an ultimatum to coal operators in the summer of 1917. In addition to union recognition, miners demanded reinstatement of discharged miners, an eight-hour day, a wage increase, union-approved checkweighmen, semimonthly payments in cash, and elimination of the subcontract system. Failure to grant the demands would result in a strike scheduled to begin on August 18. At this juncture, Secretary of Labor William B. Wilson appealed to the miners' sense of patriotism and concern for issues of national security and convinced the UMWA to call off the strike. This settlement proved temporary, but miners did receive guarantees involving union membership, grievance procedures, checkweighmen, and semimonthly paydays.[39]

Subsequently, Dr. Harry Garfield, head of the federal Fuel Administration, prepared a more permanent proposal. Known as the Garfield Agreement, this document granted miners the right to join the union, to reinstate blacklisted workers, to assemble freely, to adjudicate disputes, to appoint checkweighmen, and to receive semimonthly pay. The omission of the eight-hour day prompted UMWA officials to reiterate their strike threat, and Garfield added that provision to an amended version released in December. In essence, miners realized all of their demands except for union recognition. In spite of federal involvement, ACOA members resisted compliance with the Garfield Agreement and attempted to institute a virtual nationwide open shop dubbed the "American Plan." Furthermore, operators agreed to discuss issues with government representatives, but refused to negotiate with union officials.[40]

Small strikes ensued in the following year as miners attempted to compensate for increased living expenses. At Blocton, a one-week strike in February 1918 protested poor working conditions and sought relief from rising costs. Operators and public opinion viewed such action as unpatriotic and illegal during wartime conditions, and miners returned to work when faced with the threat of a government takeover.[41]

Unrest persisted throughout the remainder of the war, and the UMWA mobilized fifteen thousand Alabama miners to strike on November 1, 1919. Union officials demanded wage increases that had been deferred during the war, but operators ignored their demands and discharged numerous miners affiliated with the union. Federal agencies attempted to arbitrate, but they failed to intervene directly. Consequently, Governor Thomas Kilby deployed six National Guard companies to the strike zone, and union leaders were labeled as "unpatriotic." At Blocton, TCI officials fired Harry Adams, president of the local union, and ordered him to vacate company housing within forty-eight hours and to withdraw his children from the company school. Threatening prosecution if he failed to comply, the company "blackballed" Adams as a union member and leader.[42]

Governor Kilby deployed National Guard units to central Alabama again in the fall of 1920. At the direction of UMWA president John L. Lewis, coal miners struck on September 7 "to compel the operators to meet . . . and settle the questions at issue between the mine workers and the operators." Releasing tensions that had built for more than a decade, miners and company officials prepared for Alabama's "most stupendous confrontation between labor and capital."[43]

Recognizing that more than three-fourths of Alabama miners were black, operators resorted once again to accusations of unionist aspirations for "social equality." They imported strikebreakers, evicted miners and families, blacklisted strikers, courted public opinion, and maintained production. Sometimes company and government officials extended their intimidation tactics to include those indirectly related to the mines. For example, Mamie Day, wife of independent merchant E. C. Day, complained to her sister about nighttime activities near Piper. She wrote: "A few nights ago a car came up to our gate and . . . [the Governor's men came] in the yard with their guns and ask[ed] . . . if the miners were having a meeting." Questioning the legality of such action, she added: "I think we should have the right to tell them to stay off our dirt when we know there is no disorder going on."[44]

As company officials prepared to move strikebreakers into Marvel, an attorney for owner Ben Roden petitioned Governor Kilby to remove a local prounion deputy sheriff. Several days later, company representatives summoned the deputy to the mining office where he was seized, disarmed, threatened, and ordered to leave the community. Roden also commented on the antagonistic relationship between strikers and scabs: "Strikers at Marvel did everything they could to stop the scabs. They would meet the trains with a mob of one hundred and fifty or two hundred men . . . every morning and night, and any men who would come in for work they would follow them around, just like they were some king or side show exhibit, and jeer at them, and talk to them; and we just had to guard those men day and night." In other company towns, miners were compelled to inform military officers beforehand of any meetings consisting of more than three miners. This directive applied to union meetings and to religious services and fraternal meetings as well.[45]

Eventually, courts ruled on various appeals and upheld the companies' eviction notices. Movement from company housing to tent cities occurred in late January just as harsh winter weather set in. The UMWA donated tents and $70,000, but forty thousand displaced family members soon consumed those resources. Blacklisted and denied access to the company store, miners suffered in the February cold. Terminating the strike on February 22, union officials agreed to arbitration by Governor Kilby. Rendering his decision in mid-March, Kilby declared the strike "illegal and immoral" and supported the operators on every count. He granted neither union recognition nor wage increases, left the subcontract system intact, and refused to reinstate striking miners. Decimated by the governor's decision, the UMWA vanished again from the Alabama mining landscape. Miners, left with few options, generally opted to migrate northward to other coal fields. These dynamics created a ripple effect that left a lasting mark on central Alabama communities.[46]

The strike of 1920–1921 sealed the fate of the UMWA in Alabama's coal fields. The "roller coaster" ride toward unionization was characterized by membership surges in 1890, 1894, 1904, 1908, and 1920 and by precipitous declines following union losses. The coal-mining industry displayed a working-class solidarity unique within the industrial complex. Yet, continual failure in achieving recognition and collective action plagued the union. In 1928, UMWA officials closed the District 20 headquarters. For more than a decade, union activities in Alabama's coal fields lay dormant.[47]

Finally, the National Industrial Recovery Act (1933) and the National Labor Relations Act (1935) resuscitated Alabama's UMWA. Invigorated by New Deal legislation and with federal and state support, the miners' union resurrected District 20. Organizing 100 percent of Alabama's miners within two months, the UMWA launched a vigorous labor movement that successfully challenged total operator control. Confrontations at Piper-Coleanor and at Margaret epitomized the hard-fought (and often violent) struggle that ensued as prounion forces moved toward recognition. UMWA's national president, John L. Lewis, perfected the strike, eliminated nonunion mines, and supported standardized freight rates. According to labor historians Edwin Brown and Colin Davis, "From 1935, industrial relations in Alabama coal mining became less dramatic . . . because the district was well organized." Ultimately, the New Deal, the Congress of Industrial Organizations (CIO), and World War II would boost union activity to preeminence. Also, the UMWA would become the exclusive bargaining agent for coal miners until the Taft-Hartley Act levied restrictions in 1947.[48]

On the other hand, Marlene Rikard suggests that New Deal legislation created a tripartite arrangement that excluded rank-and-file coal miners. Having gained legitimacy, big labor joined big business and big government to exploit the common worker. For example, John L. Lewis supported mechanization of the coal-mining industry even though it cost many miners their jobs. Similarly, Crandall Shifflett contends that some miners believed that they had exchanged one master for another. Nevertheless, he asserts that independent miners rationally and effectively defended their interests in dealing with operators and union officials. In sum, they subscribed to a universal concept that "both management and labor should work together for the good of all."[49]

8
DECLINE AND DEMISE
1929–1976

Elizabeth Lowery married Edgar Frost, a trained barber, around 1927, and the couple moved to Dogwood in Shelby County. With a father who worked the mines at Aldrich, Lizzie felt no hesitation when Edgar began work as a miner at the Little Gem Coal Company. After two years in the mining camp, Edgar and Lizzie moved to her family's land located about halfway between Montevallo and Calera. Edgar ate breakfast each day at 2:00 A.M. before riding his mule or horse to work. If the animal threw a shoe, Edgar walked the five miles to ensure that he arrived at the mine opening before the mantrip departed at 7:00. Returning home between 7:00 and 9:00 P.M., he repeated the routine three or four times per week.

By 1945, Edgar operated a coal-cutting machine and netted about $45 per two-week pay period after stoppages. His wages for that year totaled $1,855.83 with $45.10 withheld for federal tax, and union dues amounted to $1.50 per month. Other stoppages included school and doctor fees even though the Frosts resided outside of the mining community. Lizzie shopped in Calera and Montevallo, so the couple reaped no real benefit from the company's paternalism. Unfortunately, when a falling timber struck him in the chest in 1947, Edgar was forced to quit work. Although disabled and diagnosed with black lung, no health care or medical payments materialized. Apparently, the company accepted no responsibility for his injury, and union officials offered no help either. Later, after Edgar's death in 1968, union representatives claimed that, because Lizzie was not physically present when Edgar paid his dues, she could not collect any pension payments. As she relates, the union "treated me dirty." Fortunately, the couple relied on their farm for income after Edgar's accident.[1]

Other mining families experienced similar circumstances. For example, an official study conducted by the U.S. Coal Commission (1922–1923) reported that the median annual income for a family of four equaled $930 for full-time employees and $800 for workers paid by the day. Full-time miners were paid by the ton, but they purchased their own tools, powder, and timbering materials. During the Depression, miners at Hill Creek, Belle Ellen, and Klondyke pursued "idle-work"—digging coal on "off days" to maximize production on official "work days." At the same time, the commission calculated that annual costs for food and clothing averaged $724 and $167, respectively. Thus, coal mining became a family occupation. Boys often started sorting slate at a young age, but they entered the mines as soon as practicable. Also, companies frequently hired boys (ages twelve to fourteen) who proved themselves strong enough to shovel coal and sturdy enough to endure a ten-hour day. By 1920, 1,384 children (ages ten to seventeen) mined coal in Alabama.[2]

Life was hard for coal miners like Edgar Frost and their families. Ironically, union recognition and collective bargaining arrived at the same time as a reversal in the fortunes of the coal industry. The Great Depression, market changes, and technological factors precipitated three decades of decline from the late 1920s to the mid-1950s. Shifflett considers this paradox when he observes that "the miners' hour of triumph in the late 1930s was also the moment of their greatest loss." In effect, the combined efforts of capital and organized labor resulted in unemployed miners.[3]

Historical perspective reveals that downturns in coal markets actually began before the Great Depression. The industry experienced steady growth for the first quarter of the twentieth century, but employment declined after 1927. Fortunately, miners habitually supplemented their industrial pursuits with agricultural production. Many families survived the Depression by maintaining gardens and canning fruits and vegetables. In many ways, miners reverted to the rural traditions of mutual aid and community assistance instead of relying solely on company largesse. As Shifflett reports, "company paternalism continued, but, only as a temporary expedient in dealing with an economic crisis, not as a means to attract and hold labor."[4]

Piper historian James Walker reports that peoples' lives were filled with poverty, hunger, disaster, and uncertainty. He states: "Poor working and living conditions were taken for granted by the miner, and were, in many cases, the causes of illness and disease." Families ate garden vegetables, churned butter, drank buttermilk, and obtained eggs and meat from their chickens. Peddlers plied the streets in wagons or buggies, selling beef, fish, and other food items. Walker also mentions a group that formed the lowest socioeconomic stratum—poor, uneducated people who drifted from one town to another. Meanwhile, company officials continued to live on different streets and in better houses, in an area frequently called "Silk Stocking Row."[5]

In addition to facing economic straits, coal miners endured daily hazards in the mines. A report from the U.S. Coal Mines Administration (1923) asserted that, "although mine explosions attract wide attention owing to the large number of men they involve, the deaths underground from this cause are only 12 per cent of the total." In fact, explosions ranked third among all working hazards associated with bituminous coal mining. Incidents involving underground transportation ranked second, but falling rock or coal accounted for more deaths than all other causes combined. In Alabama, the number of fatal accidents increased significantly as the mining population expanded. For example, 40 deaths in 1899 escalated to 129 two decades later, with annual fatalities peaking at 185 in 1904. Reginald Dawson's reforms and the Mine Inspection Act of 1890 established regular inspections for mines employing more than twenty workers. These regulations also provided for adjudication by a probate judge when companies failed to make appropriate corrections within a "reasonable time." Other Progressive-era reform measures, strengthened in 1911, called for standardized scales, adequate ventilation, emergency equipment, and escapeways. Legislation also provided procedures for preventing gas buildup, prohibited mining work by women, and restricted the young male population to those over fourteen years of age. Later, the Workman's Compensation Law of 1919 provided that in the case of death or injury, "the surviving spouse and next of kin shall receive compensation by way of damages from the employer."[6]

Alabama's accident and fatality rate exceeded the national average for the ten years following World War I. The next fifteen years showed some improvement in safety measures, but Alabama operations remained third in national rankings of the most dangerous mines. The Coal Mines Administration report argued that mining officials focused on economic issues rather than on safety procedures and that "many [operators] do not realize that in the long run better safety conditions would mean decided reduction in compensation payments." One exception to the rule involves Charles F. DeBardeleben's Alabama Fuel and Iron Company (AF&IC). Even though explosions occurred at Acton (1913), Margaret (1914), and Overton (1925), DeBardeleben made safety a priority. After issuing a book of safety regulations to each miner in 1929, DeBardeleben used a three-tiered policy in dealing with violations. The first offense garnered a warning, the second, suspension, and the third, discharge. Drunkenness constituted grounds for immediate discharge. Furthermore, inspectors checked the mines regularly, and foremen inspected workers for matches, pipes, and cigarettes. Violations in these cases warranted a $5 fine. On a more positive note, the AF&IC presented awards to mines that reported no accidents during a specified period.[7]

Other companies did not prove as conscientious as DeBardeleben's AF&IC. Harry Fullman hurt his foot in a mining accident at Piper in 1927. As he stated, "They brought me out of the mine . . . and dumped me on the porch." After lying alone for most of the day, Fullman eventually received help from a fellow miner who put him on the train to Birmingham. At Norwood Hospital, doctors detected gangrene and amputated Fullman's foot. The company paid 96 cents per day in compensation, and Fullman worked at a local sawmill until he had saved $131, the price for an artificial leg. He then returned to the mines, working an additional thirty years at Marvel and Piper. He averaged 11.5 tons per day and became president of the United Mine Workers organization at Marvel. Later, when Fullman's friends encouraged him to run for the Bibb County Commission, Marvel owner B. F. Roden handed him $50 to cover the qualifying fee.[8]

Accidents did not constitute the only hazards faced by coal miners. The cool, damp atmosphere caused miners to be more susceptible to tuberculosis, influenza, and pneumonia. The lack of exposure to sunshine often extended the healing time required for sores and abrasions. Furthermore, the fine dust generated by the mining process settled in the miners' respiratory systems and later manifested itself as "black lung." The Coal Mines Administration determined to study the following problems inherent in the bituminous coal mining industry:

> What conditions will yield plenty of coal at the lowest price consistent with decent living for the workers and fair profits for the owners, not merely in 1924 but in the years to come?
>
> 1. Are the workers getting a "decent living"?
> 2. Are the owners getting a "fair profit"?
> 3. How can the country be assured of "plenty of coal," and why has there sometimes been shortage instead of plenty?
> 4. Can the price of coal be lowered by reducing an element of cost or profit, and still yield "decent living" and "fair profit"?
>
> Is our coal being mined with reasonable regard for the supply available in years to come?

The solution to balancing "decent living" and "fair profit" emerged in the form of technological advancements and mechanization.[9]

Technology affected the coal industry directly in a fundamental shift reminiscent of the move from wood to coal in 1885. By the 1920s, coal faced competition from other fuel sources. In fact, from 1913 to 1925, consumption rates for natural gas increased 104 percent. Demand for petroleum products soared 211 percent, and hydroelectric power usage expanded by 119 percent. In contrast, coal consumption increased by a mere 3.3 percent for the same time frame. During the Depression, only steam or slack coal remained in relatively high demand; therefore, the Little Cahaba Coal Company at Piper sold its lump coal for $1 per ton or crushed the lumps to make slack for use in manufacturing plants, locomotives, and steamships. Consequently, coal operators sought to produce more coal with less labor and thereby lower the costs of production.[10]

Eventually, quantity outpaced quality production. As early as the 1870s, a steam-powered puncher tool hammered into coal at a rate exceeding a hundred times per minute. Shortly thereafter, an electrically powered coal-cutting machine replaced the steam-powered puncher. At the turn of the twentieth century, 25 percent of bituminous coal operations used the cutting machine; by 1920, two-thirds of bituminous output was undercut mechanically. The mechanical loader complemented the coal cutter in the early 1920s, and the continuous miner replaced both machines in the late 1940s. This revolving drum containing sharp teeth cut into the coal face along an extended front and dumped the coal onto a conveyor belt. With this innovation, a three-man crew performed the work of seventy hand-loaders, and pay rates changed from piecework based on tonnage to standard wages computed on an eight-hour day.[11]

Mechanization also introduced different mining methods. For example, the continuous miner facilitated long-wall mining, or the extraction of coal along a lengthy front. Eliminating the need for tunnels and pillars, this technique removed virtually all coal from the seam. Representing the underground miners' answer to strip mining, the long-wall method produced significant quantities of coal. On the other hand, the speed of the continuous miner frequently generated large amounts of methane and resulted in poor visibility due to increased quantities of coal dust. The fine coal dust also contributed to more cases of black lung.[12]

In similar fashion, the short-wall method worked a large portion of the coal face, but limited the span to a 150-foot front. This technique proved slightly slower and less efficient than the long-wall method, but miners found it safer. A less expensive method used a large bit (approximately five feet in diameter) to bore horizontally into a coal seam. A relatively small workforce extracted large quantities of coal, but this technique proved both wasteful and environmentally messy.[13]

As mechanization developed further, surface mining replaced underground mining as the preferred method of extraction. Strip mining, a variation of the open pit process used initially in the mid-nineteenth century, used bucket-wheel excavators to remove up to two hundred cubic yards of dirt and rock in one bucketful. Volume production, relatively low labor costs, and improved safety records contributed to surface mining's profitability and success. On the other hand, stream silting, water pollution from sulfuric acid, and soil erosion constituted significant environmental problems. Moreover, the vast scope of strip mining resulted in deforestation and reconfigured landscapes.[14]

Strip mining affected the Cahaba coal field in the Blocton area, but underground mining characterized most operations within the region. Walking or riding to the coal face, miners in the Cahaba field generally used the room and pillar method of extraction. Mechanization altered the techniques and increased production, but improved technology also resulted in fewer jobs for hand-loading miners. In 1940, two-thirds of Alabama's coal production resulted from traditional blasting and hand-loading methods. However, the introduction of mining machinery significantly reduced that percentage, and the mining population plummeted. Trade unions had gained legitimacy in the previous decade, but promises of higher wages, improved benefits, safer mines, and greater autonomy seemed hollow ten years later. Shifflett concludes that "the greatest failure of unionism . . . was the impoverishment of thousands of black and white miners who were thrust into the maw of mechanization."[15]

In addition to reducing the need for miners, machinery also affected the demographics of the coal industry. World War I reduced the immigrant population among Alabama's miners, and black laborers endured the negative repercussions of mechanization. By 1960, native-born whites comprised 90 percent of the state's mining population. Furthermore, mechanization reduced the sense of social fraternity among miners and often curtailed their freedom and independence. Also, machinery created greater quantities of fine coal dust, thus precipitating an increase in black lung cases throughout the 1940s and 1950s.[16]

The most significant technological impact on the coal industry stemmed from factors outside the mines. The automobile revolutionized the lives of coal miners, allowing many second-generation miners to move away from the company towns and to commute to work. Reducing travel time between Blocton and Birmingham to about two hours, the automobile replaced the train as the primary source of entertainment and escape. Special-order vehicles made their debut in West Blocton around 1911, and new city ordinances required front and back lights, set an eight-mile-per-hour speed limit, and mandated three days' labor from every citizen for annual road upkeep and repair. By the 1920s, Main Street looked both anachronistic and futuristic as it combined automobile traffic with hitching posts and watering troughs. However, a proposal to pave both Main and School Streets met defeat in 1925 due to the "poor industrial outlook."[17]

Disaster struck West Blocton two years later when a spark from a malfunctioning electric motor ignited a vat of cleaning fluid at Pope Brothers Cleaning and Pressing Shop. By the end of the day on July 12, 1927, the town's Main Street was devastated. Forty stores, several offices, and twenty residences succumbed to the blaze with losses totaling $500,000. Some uninsured or poorly insured merchants elected not to rebuild, but approximately three-fourths of the decimated businesses participated in a community-wide revitalization program. West Blocton formed a Chamber of Commerce and actively pursued diversified agriculture, hard-surfaced roads, extended trade territories, and industrial development geared toward mining enterprises. This resurgence was short-lived, however, as Tennessee Coal, Iron, and Railroad Company (TCI) determined to close its Blocton operations in 1928.[18]

Basing their decision on Blocton's relatively remote location, TCI leaders decided to concentrate their efforts on operations based at Ensley and Fairfield. As TCI withdrew from the area, the Blocton Red Ash Coal Company assumed control of Number 10 Mine. President W. R. Young maintained the company school, but the smaller-scale operations forced many office workers, teachers, and prominent citizens to move. The stock market crash of 1929 exacerbated this economic and civic decline, and the Blocton area experienced the full force of the Depression. As Adams relates, events of 1930 tolled "the final death knell for old Blocton, the company town. . . . Markets are down, mines practically idle, unemployment is general and the best we can do is to hope for better times." Later that year, the Mobile & Ohio Railroad terminated operations between Eoline and Blocton, and the Blocton Red Ash Coal Company closed Number 10 Mine due to expenses incurred by continuous pumping operations. The property then reverted to TCI control, and the former economic mainstay dismantled its offices, company store, and hospital. Workers removed every structure except residences by January 1931, and the Moffat mine closed the following month. Rail traffic dwindled to one train per day, and passenger service finally ended in November 1932.[19]

World War II provided some stimulus to coal markets, but the industry continued its general decline. The 1950s bode misfortune for many communities as mine closings became widespread. For example, Hill Creek Mine and Marvel Number 3 Mine closed in 1950. The following year, the Roden Coal Company announced its closure of the remaining Marvel mines, and all operations ceased in April 1954. Boothton terminated its operations in October 1950 when TCI canceled its contract with the Black Diamond Company at West Blocton. Company officials then sold all houses, warehouses, mining equipment, church buildings, stores, and schools. The Bessemer, Coal, Iron, and Land Company closed Belle Ellen in the spring of 1950, and only sixteen families remained in that community by 1953.[20]

Piper and Coleanor experienced a similar fate. Piper Number 1 closed in the mid-1930s after the top house burned. The Little Cahaba Coal Company rebuilt the structure, but the shaft had deteriorated in the interim, thereby creating unsafe conditions. Thus, as former resident Walter Gardner reports: "The decision to abandon the mine was made after the management determined the approximate cost to clean up the slope." When the Coleanor mine closed in 1945, only Piper Number 2 remained in operation. Shortly thereafter, company officials determined that competition from petroleum and natural gas, rising labor costs, and increasing mining depths rendered their operations impractical.[21]

However, Calvin Jones purchased the mines in 1946 and opened Piper Number 3 located between Numbers 1 and 2. The mines changed hands again in 1947 when Temple Tutwiler II and Joe Shaw formed the Shaw-Tutwiler Corporation. These partners leased the land between Coleanor and Marvel and contracted with the Ben Hardaway Construction Company to strip-mine that area. Meanwhile, they moved pumps and other equipment from Piper Number 2 to Piper Number 3 and hauled their coal to the Coleanor tipple. This enterprise employed approximately forty-five workers until fire broke out in Piper Number 2 in the early 1950s. Workers filled the mine with water pumped from the Cahaba River to extinguish the flames, but the hot coal seams reignited when water levels were reduced. Consequently, operators decided to seal the mine to prevent a possible explosion. At that time, company officials sold all houses and buildings at the rate of $100 per room. Nearly all of the structures were dismantled for reconstruction elsewhere, and all mining operations ended in 1954.[22]

As mining communities closed throughout the Cahaba coal field, many small-scale entrepreneurs opened "truck mines" or "wagon mines." A personal account collected by the Shelby County Historical Society relates the experience of Oscar Harrison as told by miner John Gates of Helena. Harrison and his sons "robbed" old mines by salvaging coal from pillars and rock falls. They constructed a tipple and coal chute from pine poles and laid rails into the mine. Using an old automobile engine and a series of pulleys, they extracted two coal cars at a time along the track that "threatened to fall apart every time it was touched." Daily output varied, but wagon mining spread throughout the Cahaba field. Each effort's lifespan depended on the availability and duration of the mine's coal and the miner's cash. Gates likened truck mining to lightning bugs: "they flash up and then fade out."[23]

Vestavia Hills resident and local historian Virgil Rice confirms the nature of Gates's account. Rice, a long-time employee in industrial sales with Alabama Power Company, handled contracts, complaints, and collection among these truck miners. His account reveals that these scavengers faced many of the same hazards as full-time miners. They operated "on a shoestring," relying on a tipple but had no permanent office. Running electrical wires through the woods, they frequently used soft-drink bottles as insulators. Competing miners often attempted to steal materials such as copper wire from each other, so someone usually stayed on the property for security. By keeping the total workforce—miners and top workers—under twenty, these pseudocapitalists circumvented laws requiring inspections of larger mines. Often, a centrally located weigh station or scale house served several mines. TCI owned the mineral rights for most of this territory, but truck miners sold their coal to apartment complexes or laundries that used steam boilers for heating.[24]

As underground operations decreased, these truck mines netted an average of 116,752 tons per year from 1936 to 1964. An average of twenty-nine mines operated each year, and the number of independent operations peaked at forty in both

1948 and 1955. From 1965 to 1972, underground mining declined from eighteen companies employing 134 workers to a single company with 6 employees. Still, underground operations produced an annual average of 25,043 tons during that eight-year period. Supplementing and overshadowing the truck and underground ventures, strip mining averaged seven operations per year from 1965 to 1991. In 1971, twelve companies located in Bibb, Jefferson, and Shelby Counties generated a combined output of 1,187,865 tons, and strip operations averaged 395,446 tons per year. Moreover, two Birmingham-based commercial operations developed in the Cahaba field. The Black Diamond Coal Mining Company operated Blocton Number 11 from 1965 to 1970, averaging 30 workers and 41,726 tons per year. Also, SEGCO Number 2 at Maylene—a captive mine for the Southern Electric Generating Company—averaged 241 workers and 512,927 tons from 1965 to 1971.[25]

Other Cahaba communities endured mine closings, but some did not vanish as many of the others did. For example, Aldrich shipped its last load of coal on July 5, 1942, after owner Durias Thomas declared that "the coal seam is quite well depleted." Upon dissolution of the Montevallo Coal Mining Company in August 1943, Thomas sold the company houses to residents at reasonable prices. Former Aldrich miners then commuted to other mines at Boothton and Marvel. Similarly, the AF&IC closed its Margaret mines in 1950 and donated the houses and church buildings to the community. Many residents stayed and incorporated the town of Margaret in 1959. The Acton mines were sold to brothers Bryan, Arthur, and George Chace in 1945, and their family mined coal for eighteen years before selling the property to the Harbert Corporation in 1968.[26]

West Blocton represents the most resilient mining community of the Cahaba field. Established as the commercial center for Blocton, Marvel, Coleanor, Piper, Belle Ellen, Lucile, and Moffat, this town in northern Bibb County adjusted well to its change of fortune. Referring to "Bibb's Magic City," county historian Rhoda Ellison remarks that, "unlike most of the other communities that grew up around the mines, [West Blocton] had the vitality to continue on its own power . . . when the coal-mining era had passed." In fact, West Blocton was not devastated by the Blocton closings in 1930, because the Hill Creek and Klondyke mines continued operations. Still, the early Depression years hit hard, and the West Blocton Savings Bank failed in October 1931. In addition, the Chamber of Commerce folded, several fraternal organizations closed, and some churches disbanded. The community lost its fire truck to foreclosure, the local newspaper ceased publication, and residential streetlights were removed.[27]

Relief came with the New Deal. The Public Works Administration began construction of a new high school building in 1934. Construction workers used lumber from the Number 1 Blocton TCI school and from the church building in Number 1 Town. Also, bricks from the abandoned coke ovens were used in the foundation, and pieces of the Number 1 school sidewalks formed a mosaic walk for the new school. The heating "system" relied on a dozen pot-bellied coal stoves in order to help the local economy, and the Works Progress Administration added a lighted athletic field in 1937.[28]

In 1938, the Black Diamond Coal Company of Birmingham reopened the TCI Number 9 Mine and hired 250 employees. Opening a store complete with groceries, dry goods, and furniture, the company owned the largest and busiest enterprise in town. Wages remained low due to the depressed economy, but local mines operated at full capacity. Increased business revitalized the Main Street area, and the 1940 census revealed a small increase in population.[29]

The 1940s proved to be a volatile decade as strikes and fires took their toll. Fire destroyed the tipple, coal washer, and two rail cars at the Klondyke mine in May 1941, and operators decided to remain closed due to inadequate insurance coverage. World War II interrupted mining operations for a time as many workers left for military service. Most of those miners returned to West Blocton after the war, and the Black Diamond Company expanded its operations. Fire ravaged Main Street again in January 1948, but store owners reopened in spite of warnings that "the price of coal has about reached the place where it is entirely out of the reach of the common man." Decreased demand stopped mining operations in the spring of 1948, and West Blocton experienced yet another decline in its economic fortunes.[30]

As other mining operations closed in the early 1950s, West Blocton struggled to survive. The Black Diamond Company reduced its operations in May 1951 and then closed the Number 9 Mine three years later. Also, "the long-standing reputation of the town's organized labor troubles [was] a major deterrence to attracting new industry." In spite of these drawbacks, the population of West Blocton remained relatively stable. Increased mobility allowed residents to commute to jobs in Tuscaloosa, Centreville, Bessemer, and Birmingham. Nevertheless, the coal-mining industry reached a low ebb. In 1956, only seventy-five mining jobs existed in all of Bibb County, and the 1960 census revealed a population decrease of nearly 10,000 since the "glory days" of the 1920s.[31]

The coal industry's "last gasp" came in 1963 when local surface mining operations rejuvenated the Black Diamond Company's coal washer and switch engines. Also, the Black Diamond Company opened the Number 11 slope in the vicinity of the old TCI Number 3 Mine at Smith Hill. A new coal preparation plant promised new jobs as well. A decade later, environmental issues and a contract cancellation by U.S. Steel resulted in the closure of the Black Diamond Company and the Burgess Surface Mining Company. By the end of the summer of 1976, all mining operations ceased. In the early 1980s, descriptions of West Blocton included boarded-up stores, 20 percent unemployment, homes without running water, and depressed conditions akin to regions of Appalachia.[32]

In the final analysis, approximately 150 mines operated among the coal seams that stretched along the Cahaba River Valley. Each mining community contributed its unique qualities to the flavor of the whole, but, ultimately, the Cahaba field followed the path of other Appalachian coal regions. Whether examining the poignant histories of communities like West Blocton, Aldrich, and Margaret, or reminiscing about life in the ghost towns of Piper, Marvel, and Straven, the Cahaba coal field seems to echo the names of once familiar mines and their company towns—Acmar, Acton, Aldrich, Belle Ellen, Blocton, Boothton, Climax (Maylene), Coalmont, Coleanor, Continental (Marvel), Dixie (Moffat), Dogwood, Eureka, Garnsey, Glen Carbon, Hargrove, Helena, Henry Ellen (Leeds), Hill Creek, Klondyke, Lucile, Margaret, Marvel, Montevallo, Mossboro (Helena), New Acton, Norris, Overton, Paramount, Peerless (Straven), Piper, Ratliff (Irondale), Red Diamond, Red Eagle, Riley (Leeds), Robinson (Overton), Shades Valley (Irondale), Sicard, Straven.[33]

Experiencing both boom and bust, the coal industry within the Cahaba field affected the lives of thousands of people in central Alabama. Often forgotten as they now lie dormant along the Cahaba River basin, the coal mines of the Cahaba field made a lasting impression on the families who developed and worked them. The story of Cahaba coal rests in the hearts of the people it touched.

NOTES

Introduction

1. Captive mines were affiliated directly with iron- and steel-making ventures. In this case, Warrior coal was mined exclusively for use within the Birmingham District. Alabama's coal deposits lie in four basins or fields. The Warrior field, Alabama's largest, contains 2,600 square miles with principal production in Jefferson, Walker, and Tuscaloosa Counties. The Cahaba field encompasses 270 square miles underlying Shelby, Jefferson, St. Clair, and Bibb Counties. The Coosa Field lies adjacent to and east of the Cahaba field, containing 345 square miles in Shelby and St. Clair Counties. Finally, the Blount Mountain Basin or Plateau field includes 100 square miles in Blount and Etowah Counties. State of Alabama, *Annual Report of Coal Mines* (Birmingham, AL: Dispatch Printing Co., 1928), 20–22.

2. Gavin Wright, *Old South, New South: Revolutions in the Southern Economy Since the Civil War* (New York: Basic Books, 1986), 11, 18, 28. J. Mills Thornton, III, *Politics and Power in a Slave Society: Alabama, 1800–1860* (Baton Rouge: Louisiana State University Press, 1978), 280–281. Thornton qualifies his argument by admitting that "Alabama was very far from industrialization in 1860. . . . She did have an annual industrial product of $10.5 million, more than double the figure for 1850. But almost all of this output was from light industry. Indeed, 40 percent of it came from just two sources—saw mills and grist mills. These two forms of enterprise accounted for 572 of the state's 1,459 manufacturing establishments, and employed 25 percent of its industrial work force of 7,900."

3. Curtis J. Evans, *The Conquest of Labor: Daniel Pratt and Southern Industrialization* (Baton Rouge: Louisiana State University Press, 2001), 3–5, 14, 33, 271–272, 292–293, 301. For a historiographical sketch concerning industrialization in the antebellum South, see Evans, 166–169.

4. Eugene A. Smith, *Geological Survey of Alabama* (Montgomery, AL: The Brown Printing Co., 1890). For twentieth-century comparisons see Jack C. Pashin et. al., *Geology and Coal Resources of the Cahaba Coal Field,* Bulletin 163 (Tuscaloosa: Geological Survey of Alabama, 1995); Reynold Q. Shotts, *Coal Reserves of Bibb County, Alabama*, Geological Survey of Alabama, Circular 72 (University: University of Alabama, Division of Economic Geology, 1971); and *A Summary of the Composition and Technology of the Coals of Bibb County, Alabama*, Geological Survey of Alabama, Circular 73 (University: University of Alabama, Division of Economic Geology, 1971).

Chapter 1

1. Ethel Armes, *The Story of Coal and Iron in Alabama* (Birmingham, AL: The Chamber of Commerce, 1910), 10. "Cahaba" constitutes a derivative from the Choctaw "Oka aba." "Oka aba" signifies the river above, and the name may first have been given to this river by some Choctaw-speaking Indians living down on the lower course of the Alabama River.

2. Armes, *Coal and Iron,* 25.

3. Ibid., 73–74. Armes also states that "the precise locality is in Section 11, Township 21, Range 5, east, and the mines of Garnsey, operated today by the Galloway Coal Company, are located here."

4. T. H. Aldrich, "Historical Account of Coal Mining Operations in Alabama Since 1853," in Eugene A. Smith, *Geological Survey of Alabama: Report of Progress for 1875* (Montgomery: W. W. Screws, 1876), 28–29; "In the Coal Fields," *Birmingham Daily Age-Herald,* January 6, 1889.

5. Lewis S. Dean, ed., *Michael Tuomey's Reports and Letters on the Geology of Alabama, 1847–1856,* Information Series 77 (Tuscaloosa: Geological Survey of Alabama, 1995), 1.

6. Smith, *Report of Progress for 1875,* 205–212; Wallace, *St. Clair: A Nineteenth-Century Coal Town's Experience with a Disaster-Prone Industry* (New York: Alfred A. Knopf, 1987), 84. Similarly, Henry Darwin Rogers, writing a geological report in 1838, noted that mining in Pennsylvania's Pottsville region was hampered by faulting.

7. Smith, *Report of Progress for 1875,* 212.

8. Ibid., 28–31. Armes, *Coal and Iron,* 70, 149; J. Mills Thornton III, *Politics and Power in a Slave Society: Alabama, 1800–1860* (Baton Rouge: Louisiana State University Press, 1978), 290. J. Mills Thornton endorses these conclusions when he states that "the completion of the Alabama and Tennessee Rivers Railroad to Montevallo in 1853 . . . did allow the beginning of some coal mining in the Shelby area. William P. Browne, a former state representative from Mobile, opened a mine in that year, but the reluctance of owners to lease slaves for the dangerous work severely restricted his production. Still, by 1860 Shelby County was reporting more than $41,000 a year in coal production—virtually the entire output of the state."

9. Ronald L. Lewis, *Coal, Iron, and Slaves: Industrial Slavery in Maryland and Virginia, 1715–1865* (Westport, CT: Greenwood Press, 1979), 6–7, 45–46, 54. Sean Patrick Adams, *Old Dominion, Industrial Commonwealth: Coal, Politics, and Economy in Antebellum America* (Baltimore: The Johns Hopkins University Press, 2004), xii, 13. Adams "chose coal as a case study to explore larger questions of regional development in the nineteenth century and the nature of the Industrial Revolution in the United States."

10. Virginia Estella Knapp, "William Phineas Browne, A Yankee Business Man of the South" (master's thesis, University of Texas, 1948), 1–8. Knapp notes that Thomas M. Owen's biography of William Phineas Browne gives Browne's birthplace as Waltham, Massachusetts, his father's hometown. See Thomas M. Owen, *History of Alabama and Dictionary of Alabama Biography,* 4 vols. (Chicago: S. J. Clarke Publishing Company, 1921).

11. Knapp, "William Phineas Browne," 9–24.

12. Ibid., 26–44.

13. Ibid., 41–43.

14. Ibid., 44.

15. Margaret Stevens Browne, "near Montevallo," to William Phineas Browne, Mobile, January 28, 1849, Special Collections, Alabama Department of Archives and History, Montgomery, Alabama, hereafter cited as ADAH. At the time of Margaret's marriage to William Phineas, her mother was married to Richard Wood. Thus, Margaret's pining about not having a place of their own may suggest that she lived with her mother and stepfather until she and Willie built a house on her mother's property.

16. Martin Everse, "Sally's Story," *Shelby County Historical Quarterly* 14–16 (April 1988), 28–40. Rev. E. B. Teague, "Historical Sketches of Shelby County" (photocopy in the possession of the author). "Farm Building Noted," *Shelby [County] News* (Columbiana, AL), November 30, 1978. This article states: "Constructed of slave-made bricks, the cooling house was first used for cooling milk, vegetables, and other perishables. . . . Water was drawn from a nearby well and poured into a water box that fed a trough inside the building." Also, "the Sally Dickey Woods cooling house, now known as the Cleveland Garden House, has been named to the Alabama Register of Landmarks and Heritage by the Alabama Historical Commission. . . . The cooling house was constructed sometime in the mid-1800s on the farm of Sally Dickey Woods."

17. Richard Wood, Montevallo, to W. P. Browne, Mobile, February 11, 1849, ADAH; Philip J. Weaver, Selma, to W. P. Browne, Montevallo, November 12, 1849, ADAH.

18. John Strong Storrs, Montgomery, to W. P. Browne, Montevallo, January 23, 1850, ADAH; P. J. Weaver, Selma, to W. P. Browne, Montevallo, July 26, 1850, ADAH.

19. George O. Baker, New York, to W. P. Browne, August 21, 1851, ADAH; Mary Jane Allen, Montevallo, AL, to Margaret S. Browne, Middlebury, VT, August 3, 1853, ADAH; G. O. Baker to W. P. Browne, c. 1853, ADAH; Knapp, "William Phineas Browne." Knapp records that daughter Claudia's ill health and eventual death in April 1853 prolonged the Brownes' visit to the Northeast.

20. P. J. Weaver, New York, to W. P. Browne, September 25, 1853, ADAH; P. J. Weaver, Shelby Springs, to W. P. Browne, Philadelphia, October 28, 1853, ADAH.

21. R. H. Jackson, Greensboro, AL, to W. P. Browne, December 22, 1853, ADAH; Thomas B. Scott to W. P. Browne, January 11, 1854, ADAH.

22. Adams, *Old Dominion, Industrial Commonwealth,* 33–44; Ronald L. Lewis, *Coal, Iron, and Slaves,* 20.

23. P. J. Weaver, Selma, to W. P. Browne, February 9, and November 9, 1854, ADAH; H[orace] Ware to W. P. Browne, July 5, 1854, ADAH; C. Browne, Boston, to W. P. Browne, August 18, 1854, ADAH.

24. James Adams, Selma, to W. P. Browne, Montevallo, December 3, 1854, ADAH.

25. James Adams, Mobile, to W. P. Browne, April 2, 1855, ADAH.

26. Richard Wood to W. P. Browne, Selma, December 23, 1854, ADAH. "Poor transportation from the mines to shipping centers also plagued the early development of the Richmond basin." Adams, *Old Dominion, Industrial Commonwealth,* 34.

27. D. W. Sterrett, Camden, to W. P. Browne, November 11, 1854, ADAH; Margaret Browne, Montevallo, to W. P. Browne, Mobile, February 22, 1855, ADAH.

28. Richard Wood to W. P. Browne, December 12, 1854, ADAH; James Adams, Mobile, to W. P. Browne, January 21, 1855, ADAH; Margaret Browne, Montevallo, to W. P. Browne, Selma, February 23, 1855, ADAH.

29. J. Cartwright, Mobile, to W. P. Browne, May 12, 1855, ADAH.

30. R. H. Kerr, Selma, to W. P. Browne, May 18, 1855, ADAH; John H. Murphy & Co., Montgomery, to W. P. Browne, Montevallo, May 29, 1855, ADAH.

31. Kerr to Browne, May 18, 1855.

32. W. R. Bill, Selma, to W. P. Browne, June 6 and 19, 1855, ADAH.

33. Robert Smith, Selma, to W. P. Browne, Montevallo, September 3, 1855, ADAH; P. J. Weaver, Selma, to W. P. Browne, October 2, 1855, ADAH.

34. Contract witnessed by Jno. P. Morgan, Justice of the Peace, Shelby County, December 24, 1855, ADAH; P. J. Weaver, Selma, to W. P. Browne, December 28, 1855, ADAH; G. O. Baker, Selma, to W. P. Browne, January 2, 1856, ADAH.

35. J. Cartwright, Mobile, to W. P. Browne, January 16 and 23, 1856, ADAH; G. O. Baker, Selma, to W. P. Browne, January 23, 1856, ADAH. Coal is defined by its size: large lump coal—greater than 8 inches in diameter; medium lump coal—3 to 8 inches; nut coal—1 to 3 inches; steam or slack coal—smaller than 1 inch. Generally, families used larger pieces for heating their homes, and manufacturers reserved the finer grades for industrial purposes, such as producing steam, natural gas, coke, etc. Anthony F. C. Wallace categorizes Pennsylvania's anthracite coal into the following nine sizes: lump—greater than 7 inches; steamboat—5 to 7 inches; broken—2¾ to 5 inches; egg—2 to 2¾ inches; large store—1½ to 2 inches; small store—1 to 1½ inches; chestnut—⅝ to 1 inch; pea—⅜ to ⅝ inch; buckwheat—¼ to ⅜ inch. Anything smaller than one-quarter inch was waste. Wallace, *St. Clair,* 17.

36. G. O. Baker, Selma, to W. P. Browne, February 14, 1856, ADAH; J. Cartwright, Mobile, to W. P. Browne, March 12, 1856, ADAH.

37. Margaret Stevens Browne, Montevallo, to W. P. Browne, April 23, 1856; May 29, 1856; June 16, 18, and 21, 1856, ADAH. The *American Heritage Dictionary* defines *skippers* as "any of numerous butterflies of the families Hesperiidae and Megathymidae, having a hairy, mothlike body and a darting flight pattern."

38. J. Cartwright, Mobile, to W. P. Browne, October 16, 1856, ADAH.

39. G. O. Baker, Selma, to W. P. Browne, December 10 and 19, 1856, ADAH.

40. Margaret S. Browne, Montevallo, to W. P. Browne, c. fall 1856, ADAH.

41. G. O. Baker, Selma, to W. P. Browne, January 2, 1857, ADAH.

42. G. O. Baker, Selma, to W. P. Browne, January 5, 1857, ADAH.

43. G. O. Baker, Selma, to W. P. Browne, January 7, 9, and 19, 1857, ADAH. Wallace highlights similar problems in Pennsylvania's coal regions: irregular demand for coal, fluctuations in price, and transportation costs. Moreover, a shortage of capital marked the normal condition for mine owners. Operators secured capital by partnering with investors, borrowing cash, and/or delaying payments on tracts. Wallace, *St. Clair,* 83, 220.

44. G. O. Baker, Selma, to W. P. Browne, January 19 and 29, 1857, ADAH.

45. G. O. Baker, Selma, to W. P. Browne, February 2, 1857, ADAH.

46. G. O. Baker, Selma, to W. P. Browne, February 5 and 12, 1857, ADAH.

47. G. O. Baker, Selma, to W. P. Browne, February 12, 1857, ADAH.

48. G. O. Baker, Selma, to W. P. Browne, March 2, 1857, ADAH.

49. A. Saltmarsh, Cahaba, to W. P. Browne, March 5, 1857, ADAH; G. O. Baker, Selma, to W. P. Browne, March 5 and 16, 1857, ADAH.

50. G. O. Baker, Selma, to W. P. Browne, March 16 and 31, 1857, ADAH.

51. G. O. Baker, Selma, to W. P. Browne, May 13 and 20, 1857, ADAH.

52. G. O. Baker, Selma, to W. P. Browne, October 5 and 31, 1857, ADAH.

53. Frank Steward, Mobile, to W. P. Browne, Selma, January 3, 1858, ADAH; R. H. Kerr, Selma, to W. P. Browne, January 23, February 1, March 2 and 15, April 27, and May 1 and 4, 1858, ADAH.

54. Thomas K. Hooper, Uniontown, to W. P. Browne, June 17, 1858, ADAH; Eli Loveland, Marion, to W. P. Browne, June 21, 1858, ADAH; W. P. Browne, Selma, to J. T. Campbell, Montevallo, October 25, 1858, ADAH.

55. G. O. Baker, Selma, to W. P. Browne, September 23, 1858, and c. 1859, ADAH; Murdock & Co., Mobile, to W. P. Browne, Montevallo, December 12, 1858.

56. G. O. Baker, Selma, to W. P. Browne, January 7, 1859, ADAH.

57. Captain Buckley of S. A. Danby & Co., Montgomery, to W. P. Browne, December 23, 1858, ADAH; G. O. Baker, New York, to W. P. Browne, January 15, 1859, ADAH.

58. Margaret S. Browne, Montevallo, to W. P. Browne, October 29, and November 20, 1858, ADAH. According to Margaret's tally, the average number of trams equaled thirty per day in November 1858, but dropped to twenty-four per day in December 1858 and January 1859. Jasper Newton Campbell was born November 9, 1836, in Lebanon (Dekalb County), Alabama. He migrated to Shelby County as a young man and married Axie Elizabeth Broadhead on January 9, 1859. After serving in the Civil War, he returned to prospecting for coal seams in Bibb and Tuscaloosa Counties. Campbell died on June 25, 1911. *Heritage of Bibb County, Alabama* (Clanton, AL: Heritage Publishing Consultants, Inc., 1998), 82.

59. Margaret S. Browne, Montevallo, to W. P. Browne, January 11, 1859, ADAH.

60. G. O. Baker, New York, to W. P. Browne, April 17, 1859, ADAH.

61. James Burns per John Burns of Gas Office, Mobile, to W. P. Browne, June 9, 1859, ADAH; Eli Loveland, Marion, to W. P. Browne, August 11, 1859, ADAH. Loveland informed Browne of the specific arrangements.

62. J. F. Conoley, Shelby Springs, to W. P. Browne, Montevallo, July 25, and August 30, 1859, ADAH; G. O. Baker, New York, to W. P. Browne, September 17, 1859, ADAH.

63. G. O. Baker, New York, to W. P. Browne, September 22, 1859, ADAH.

Chapter 2

1. Joseph Squire, "Autobiography, 1908(?)" (photocopy of the handwritten manuscript Squire prepared for Ethel Armes in the possession of the author), 4–10. In his autobiography, Squire names the white miners present upon his arrival in Montevallo:

Alexander Anderson, John Whitehead, William Gentry, Jackson Gentry, Patrick Genett, James O'Neil, Michael Shannon, William Dodd, Michael Cleary and his sons George and Mick Cleary, Joseph Yeshic and his father, Ab Allen, Thomas Allen, Frank Petrashic, Ed Whalen, B. Bergar, Alexander Luke, Peter Burke, William Courtney, Hughey Lavell. For a separate account of Squire's influence on mining operations in central Alabama, see James Sanders Day, "Dealing in Black Diamonds: Joseph Squire and Alabama's Early Coal-Mining Operations," *Alabama Review* 64, no. 1 (January 2011), 3–29

2. Alex Lichtenstein, *Twice the Work of Free Labor: The Political Economy of Convict Labor in the New South* (London: Verso, 1996), 107. C. K. Yearley Jr., *Enterprise and Anthracite: Economics and Democracy in Schuylkill County, 1820–1875* (Baltimore: The Johns Hopkins Press, 1961), 53, 112.

3. The Irish Pit was one of several drift mines—those driven laterally into a seam—already opened west of Montevallo. Other mines along the Montevallo Seam included the Arcade Pit, Browne Pit, Dutch Pit, Whim Pit, Wood's Pit, and Fancher Pit. Another set of mines dotted the Overturned Measures along Little Mayberry Creek and tapped the Shaft, Dodd, Cooper, and Cannel Seams.

4. G. O. Baker, Selma, to W. P. Browne, January 15, 1860, Special Collections, Alabama Department of Archives and History, Montgomery, Alabama, hereafter cited as ADAH.

5. G. O. Baker, Selma, to W. P. Browne, January 20, 1860, ADAH.

6. Eli Loveland, Marion, to W. P. Browne, January 27, and February 15, 1860, ADAH.

7. G. O. Baker, Selma, to W. P. Browne, Montevallo, March 1, 1860, ADAH; Eli Loveland, Marion, to W. P. Browne, March 19, 1860, ADAH. Loveland reserved four carloads for retail sales, used one to supply Howard College, and sold the remaining two loads to J. F. Bondurant and L. N. Walthall. Margaret S. Browne, Montevallo, to W. P. Browne, Mobile, March 27, 1860, ADAH; Thomas A. Walker, Selma, to W. P. Browne, April 18, 1860.

8. G. O. Baker, Selma, to W. P. Browne, April 17, 21, and 28, 1860, ADAH.

9. G. O. Baker, Selma, to W. P. Browne, May 12, 1860, ADAH; A. Saltmarsh, Cahaba, to W. P. Browne, May 16, 1860, ADAH; Margaret S. Browne, Montevallo, to W. P. Browne, Richmond, June 11, 16, and 21, 1860, ADAH.

10. G. O. Baker, Selma, to W. P. Browne, August 3, 1860, ADAH; Eli Loveland, Marion, to W. P. Browne, August 30, 1860, ADAH; Ronald L. Lewis, *Coal, Iron, and Slaves: Industrial Slavery in Maryland and Virginia, 1715–1865* (Westport, CT: Greenwood Press, 1979), 20.

11. G. O. Baker, Selma, to W. P. Browne, August 6, 1860, ADAH; Eli Loveland, Marion, to W. P. Browne, August 30, 1860, ADAH.

12. G. O. Baker, Selma, to W. P. Browne, October 26, 1860, ADAH; R. A. Moseley, Columbiana, to W. P. Browne, August 23, 1860, ADAH; C. G. Samuel, Attorney at Law, Talladega, to W. P. Browne, November 28, 1860, ADAH; Ronald L. Lewis, *Coal, Iron, and Slaves,* 20.

13. Governor A. B. Moore, Marion, to W. P. Browne, October 24, 1860, ADAH.

14. G. O. Baker, Selma, to W. P. Browne, November 15, 1860, ADAH.

15. G. O. Baker, Selma, to W. P. Browne, November 20, and December 17, 1860, ADAH.

16. G. O. Baker, Selma, to W. P. Browne, December 31, 1860, ADAH.

17. William Warren Rogers, Robert David Ward, Leah Rawls Atkins, and Wayne Flynt, *Alabama: The History of a Deep South State* (Tuscaloosa: University of Alabama Press, 1994), 184; G. O. Baker, Selma, to W. P. Browne, February 14, 1861, ADAH.

18. G. O. Baker, Selma, to W. P. Browne, April 13, 15, 17, and 19, 1861, ADAH.

19. James R. Bennett, *Tannehill and the Growth of the Alabama Iron Industry* (McCalla: Alabama Historic Ironworks Commission, 1999), 22–26; *Alabama Blast Furnaces* (Woodward, AL: Woodward Iron Company, 1940), 97–99. Browne's eleven slaves consisted of six men, two women, one girl, and two small children. Virginia Estella Knapp, "William Phineas Browne, A Yankee Business Man of the South" (master's thesis, University of Texas, 1948), 43–71; Squire, "Autobiography," 14.

20. Squire, "Autobiography," 8. Other comments about the war permeated Squire's Day Book—

—Sun., May 5: "Read the war news. I believe that bloodshed will shortly begin."
—Thurs., June 13: "This was a general day of fasting and prayer for the success of the S[outher]n A[rmie]s."
—Wed., July 10: "This is the day the Southern Congress assembles. Got news tonight that the Washington Congress had determin'd on prosecuting the war."
—Tues., July 23: "One heard tonight of a great battle near Manassas Junction."

Rumors also abounded—

—Fri., Nov. 15: "The news came today that the rebel army was all dispers'd and taken prisoners in Kentucky. One account says 300 killed; another account 1000. They are praying and grieving about it very much here about Montevallo."
—Sat., Nov. 16: "Heard today that it was a false report about the affair in Kentucky."

Joseph Squire, "Diary of Joseph Squire for year 1861" (photocopy in the possession of the author), 10, 12, 14–15, 23. Ethel Armes, *The Story of Coal and Iron in Alabama* (Birmingham, AL: The Chamber of Commerce, 1910), 124–147; Bennett, *Tannehill,* 79–81, 127, 140–141.

21. Armes, *Coal and Iron,* 149.

22. G. O. Baker, Selma, to W. P. Browne, April 26, May 8 and 10, 1861, ADAH. Baker's figures included a base price—$4.60; freight to Selma—$1.40; shipping per ton—$0.40; freight to Mobile—$2.00; wharfage—$0.30; for a total cost of $8.70.

23. Squire, "Diary, 1861," 7–8.

24. Joseph Squire, Montevallo, to W. P. Browne, May 18 and 20, 1861, ADAH; Squire, "Diary, 1861," 17–19.

25. G. O. Baker, Selma, to W. P. Browne, July 21, 1861, ADAH. Baker informed Browne that "if it is under 8 tons, we pay for a carload (of 16,000 lbs) at $1.32 per ton or $10.50 for

the whole. If over 16,500 lbs we pay double on all over 16,000, but if between 16,000 & 16,500, then we pay pro rata for the excess. This leaves you no alternative but to take an oil cask, make it (by sawing it in two parts) so it will hold 250 or 200 lbs & measure it all." Squire, "Diary, 1861," 27.

26. G. O. Baker, Selma, to W. P. Browne, September 26, and October 5, 8, and 12, 1861, ADAH.

27. William Wirt, Atlanta, to W. P. Browne, February 4, 1862, ADAH; John W. Lapsley, Selma, to W. P. Browne, February 8, 1862, ADAH.

28. In similar fashion, Curtis Evans reports that Daniel Pratt supported the Confederacy, but complained that contracting authorities had exploited him. Ultimately, Pratt lost more than $500,000 due to wartime contracts. Curtis J. Evans, *The Conquest of Labor: Daniel Pratt and Southern Industrialization* (Baton Rouge: Louisiana State University Press, 2001), 223, 229; Julius A. Pratt, Selma, to W. P. Browne, Montevallo, May 16, 1863, ADAH.

29. G. O. Baker, Selma, to W. P. Browne, June 6 and 15, 1863, ADAH.

30. Squire, "Autobiography," 17–18.

31. William Phineas Browne, Selma, to Captain John K. Mitchell, December 3, 1862, ADAH; P. M. Fancher, Montevallo, to W. P. Browne, Selma, March 1, 1863, ADAH; W. W. Atmar, Choctaw County, AL, to W. P. Browne, January 11, 1863, ADAH. Atmar gave detailed dimensions and instructions for constructing an iron furnace. Isaac D. Spear, Mobile, to W. P. Browne, Montevallo, January 18 and July 11, 1863, and contract dated July 16, 1863, ADAH. Spear's letterhead identifies him as an "Importer and Dealer in Tin Plate, Sheet Iron, Zinc, Copper, Wire, Block Tin, Spelter, Solder, &c. Also Agent for Engines and Saw Mills, and Hoe's Circular Saws, 12 Commerce Street, Mobile." Eugene A. Smith, *Geological Survey of Alabama: Report of Progress for 1875* (Montgomery: W. W. Screws, 1876), 31.

32. G. O. Baker, Selma, to W. P. Browne, July 31, 1863, ADAH.

33. Some accounts spell Gould with a double "O," that is, "Goold." *Birmingham Daily Age-Herald,* January 6, 1889; *Birmingham News—Age-Herald,* August 10, 1932; Lewis Shannon, *Billy Gould: Coal Mine and Coke Oven Site* (National Park Service, Historic American Engineering Record, Birmingham Recording Project, 1992). After the war, Gould and his partners sold their operation to the Cahaba Coal Company. That enterprise produced 40,000 tons of coal and coke over the next five years, but ceased operations in 1870. S. D. Holt leased the mine in 1874 and produced 60 tons per day. No records indicate production after 1875.

34. Armes, *Coal and Iron,* 150–151. Armes names some of these refugees—Brooks, Gainer, Rogers, Carter, Gholson, Herndon, and Thompson. In the twentieth century, the Lower Thompson mine would be known as Piper Number Two, and the Upper Thompson mine would become Coleanor.

35. Armes, *Coal and Iron,* 161–162. John T. Milner, *Report of the Chief Engineer to the President and Board of Directors of the South and North Alabama Railroad Co., On the 26th of November, 1859* (Montgomery, AL: Advertiser Steam Printing House, 1859), 3. Milner

also supported a continuation of the route through Elyton, thereby gaining access to the resources of Jones Valley and the Warrior coal field.

36. Armes, *Coal and Iron,* 161–164. Armes identifies these shareholders as twenty-five planters and businessmen of Alabama and Mississippi, among whom, besides members of the Gilmer and Noble families of Montgomery, were B. S. Bibb, T. L. Mount, M. E. Pratt, and Daniel Pratt, Bennett assesses production as follows: "Considering the size of the stack and the highly silicious content of the red hematite ore (soft) it is doubtful whether the daily capacity of the furnace ever exceeded five or six tons. In order to produce this amount of iron, not less than 60 men were required at the furnace and between 200 and 300 slaves for cutting and hauling wood for the charcoal burners. . . . The rebuilt furnaces were put into blast late in 1873 but the anticipated increase in production was not realized, due mainly to the inexperience of the operators. Though rated at 25 tons a day each the furnaces but seldom made more than 10 tons each." *Alabama Blast Furnaces,* 107.

37. John T. Milner, Montgomery, to William B. Gilmer, Montgomery, March 17, 1864 (photocopy in the possession of the author).

38. Armes, *Coal and Iron,* 194–195. In fact, the Hale & Murdock Furnace in Lamar County remained untouched. See *Alabama Blast Furnaces,* 74.

39. G. O. Baker, Selma, to W. P. Browne, Montevallo, June 12, 1865; April 2 and 5, 1867, ADAH. Baker identifies these freemen as Robert Brooks ("the best one of the lot & may make a good leader"), William Bibb, Steward Kelly, Washington Phillips and wife Mary, Raford Bell, Green Bell, Jim Bell, Joe Bell, Dempsy Spann and wife Laura, Daniel Bell, and London Bibb. Smith, *Report of Progress for 1875,* 31.

40. J. T. Leeper, Attorney for Mobile & Selma Coal Mining Company, to W. P. Browne, Montevallo, November 4, 1867, ADAH; R. M. Moore, Selma, to W. P. Browne, Montevallo, February 1, 1868, ADAH.

41. Key officials and investors of the Montevallo Coal Mining Company included Captain T. J. Portis of Dallas County, president; George M. Figh of Montgomery, general superintendent; Benjamin B. Davis of Montgomery, secretary and bookkeeper; Rev. I. T. Tichenor and future governor Thomas H. Watts of Montgomery, stockholders. Squire, "Autobiography," 19–21. Squire reported that one bullet remained in his lower leg until May 1866. An initial operation by a Dr. Hendree failed, but a subsequent attempt by a Dr. Wadkins of Columbiana proved successful.

42. Squire, "Autobiography," 24. Squire noted that settlers throughout the Cahaba coal field believed that he was surveying for the federal government in order to give land to free blacks. In fact, threats prompted him to dismiss his black assistant and to hire white workers. Nevertheless, "their women folks would beg them to stay away from me, on account of the risk of their getting killed along with me." Squire stated that he was armed at all times. Kimball Johnson, "Isaac Taylor Tichenor: A Biography" (ThD thesis, Southern Baptist Theological Seminary, 1955), 69. See also Michael E. Williams Sr., *Isaac Taylor Tichenor: The Creation of the Baptist New South* (Tuscaloosa: University of Alabama Press, 2005), 59–69.

43. Johnson, "Isaac Taylor Tichenor," 70–71. Squire, "Autobiography," 24–25.

44. Johnson, "Isaac Taylor Tichenor," 117. Williams, *Isaac Taylor Tichenor,* 70–96.

45. Squire, "Autobiography," 26–29.

46. Obviously, this marks the initial stage in a series of events that led to the Oxmoor tests of 1876 that proved Cahaba coal inferior to Warrior coal for coking. Squire, "Autobiography," 32. Sadly, DeBardeleben's papers remain unprocessed in the Birmingham Public Library Archives. A collection of letters salvaged by Ken Penhale of Helena from Squire's decrepit barn, however, sheds light on the development of the region and the construction of the *Geological Survey.* This correspondence between Squire and Aldrich, Smith, and Tichenor provides significant detail about the exploration and exploitation of the Cahaba field from 1873 to 1890. Furthermore, Squire's practice of drafting his reply in pencil on the back of received letters provides the researcher with "both sides" of the correspondence.

47. Squire, "Autobiography," 33. The story of this complex relationship among a quartet of Alabama's most important late-nineteenth-century entrepreneurs is revealed in correspondence between Squire and Aldrich that began in September 1874. Their sporadic letters indicate a growing professional relationship.

48. Ibid., 33–35.

49. W. David Lewis, *Sloss Furnaces and the Rise of the Birmingham District: An Industrial Epic* (Tuscaloosa: University of Alabama Press, 1994), 55–58.

50. Justin Fuller, "Henry F. DeBardeleben, Industrialist of the New South," *Alabama Review* 39, no. 1 (January 1986), 3–12.

51. George Brown Tindall and David E. Shi, *America: A Narrative History,* 8th ed. (New York: W. W. Norton, 2009), 749–750; Jonathan M. Wiener, *Social Origins of the New South: Alabama, 1860–1885* (Baton Rouge: Louisiana State University Press, 1978), 209–215.

52. Squire, "Autobiography," 35–36.

53. Eugene A. Smith, Tuscaloosa, Alabama, to Joseph Squire, Helena, January 23, 1880, transcript in private collection of Ken Penhale, Helena, Alabama, hereafter cited as Penhale transcript.

54. Squire, "Autobiography," 38.

55. W. C. Thompson, Dudley, to Joseph Squire, Helena, December 10, 1882, Penhale transcript. Squire, "Autobiography," 38.

56. Truman H. Aldrich, Dudley, to Joseph Squire, Helena, January 24, 1883, and February 4, 1883, and March 5 and 11, 1883, Penhale transcript; Joseph Squire, Helena, to Truman H. Aldrich, Dudley, March 4, 12, and 14, 1883, Penhale transcript.

57. State appropriations included $800 to buy chemicals and laboratory apparatus for analysis of soils, ores, minerals, and mineral water; $2,200 to purchase an ambulance and team; and an annual sum of $500 to pay traveling and incidental expenses of the state geologist. *Acts of the [45th] Session of 1872–73 of the General Assembly of Alabama held in the City of Montgomery, commencing November 18th, 1872* (Montgomery, AL: Arthur Bingham, 1873), 89–91; Eugene A. Smith, Tuscaloosa, to Joseph Squire, Helena, April 18, 1883, Penhale transcript; Armes, *Coal and Iron,* 264.

58. Lewis S. Dean, ed. *Michael Tuomey's Reports and Letters on the Geology of Ala-*

bama, 1847–1856, Information Series 77 (Tuscaloosa: Geological Survey of Alabama, 1995), 1; Smith, *Report of Progress for 1875,* 205–212; Patsy Gerstner, *Henry Darwin Rogers, 1808–1866: American Geologist* (Tuscaloosa: University of Alabama Press, 1994), 44–47. States initiating geological surveys in the 1820s and 1830s included North Carolina (1823), South Carolina (1824), Massachusetts (1830), Tennessee (1831), Maryland (1833), New Jersey (1833), Connecticut (1835), Virginia (1835), Maine (1836), New York (1836), Ohio (1836), and Pennsylvania (1836). "By 1850 twenty states had or had had state geological surveys." Anne Millbrooke, "The Geological Society of Pennsylvania, 1832–1836, Part 1, Founding the Society," *Pennsylvania Geology* 7 (1976), 7. The state geologist of Tennessee, Gerard Troost, conducted the first unofficial survey of Alabama lands in April 1834. Gerard Troost, "The Beginning of Geological Exploration in Northern Alabama: Gerard Troost's Travels in 1834," edited by James X. Corgan and Michael A. Gibson, *Alabama Review* 48, no. 2 (April 1995), 142–143, 152. Five states—Alabama, Kentucky, Minnesota, Pennsylvania, and Wisconsin—reactivated state surveys during the national depression of the early 1870s in an attempt to promote economic development. Anne Marie Millbrooke, "State Geological Surveys of the Nineteenth Century" (PhD diss., University of Pennsylvania, 1981), 122.

59. Sir Charles Lyell, *A Second Visit to the United States of North America* (New York: Harper & Brothers, 1849), 67, 70; "'Coal Field of Tuscaloosa, Alabama'—being an extract of a letter to Prof. Silliman, from Charles Lyell, Esq., dated Mobile, Alabama, Feb. 19th, 1846," *The American Journal of Science and Arts* 1 (May 1846), 375; Thomas McAdory Owen and Marie Bankhead Owen, *History of Alabama and Dictionary of Alabama Biography,* vol. 3 (Chicago: The S. J. Clarke Publishing Company, 1921), 240.

60. State appropriations provided $5,000 per year for the next ten years: that is, the state geologist's salary was budgeted at $200 per month, not to exceed $1,000 per year, and $4,000 for operating expenses. *Acts of the [52d] General Assembly of Alabama, passed at the Session of 1882–83, held in The City of Montgomery, commencing 2d Tuesday in November 1882* (Montgomery, AL: W. D. Brown & Co., 1883), 57–59; Squire, "Autobiography," 51. Delays experienced by Smith and Squire were typical; see Millbrooke, "State Geological Surveys of the Nineteenth Century."

Chapter 3

1. Eugene A. Smith, Tuscaloosa, to Joseph Squire, Helena, August 9, 1883, transcript in private collection of Ken Penhale, Helena, Alabama, hereafter cited as Penhale transcript.

2. Eugene A. Smith, Tuscaloosa, to Joseph Squire, Helena, January 12, and April 29, 1884, Penhale transcript; Joseph Squire, Helena, to Eugene A. Smith, Tuscaloosa, January 19, and February 11, 1884, Penhale transcript; Joseph Squire, Helena, to Henry A. McCalley, Birmingham, April 19, 1884, Penhale transcript.

3. Eugene A. Smith, Tuscaloosa, to Joseph Squire, Helena, May 19, 1884, Penhale transcript; Joseph Squire, Helena, to Eugene A. Smith, Tuscaloosa, June 3, 1885, Penhale transcript.

4. Joseph Squire, Helena, to Eugene A. Smith, Tuscaloosa, October 12, 1885, Pen-

hale transcript. Because the Tichenor correspondence begins in December 1885, it seems logical that "Atlanta parties" meant Tichenor.

5. I. T. Tichenor, Atlanta, Georgia, to Joseph Squire, Helena, Alabama, January 23, 1886, Penhale transcript.

6. I. T. Tichenor, Atlanta, to Joseph Squire, Helena, June 1, 1886, Penhale transcript.

7. I. T. Tichenor, Atlanta, to Joseph Squire, Helena, June 4 and 9, 1886, Penhale transcript.

8. I. T. Tichenor, Atlanta, to Joseph Squire, Helena, June 14, and July 24, 1886, Penhale transcript.

9. I. T. Tichenor, Atlanta, to Joseph Squire, Helena, October 28, and November 1, 1886, Penhale transcript.

10. I. T. Tichenor, Atlanta, to Joseph Squire, Helena, November 11, 1886, Penhale transcript; Joseph Squire, Helena, to I. T. Tichenor, Atlanta, November 20, 1886, Penhale transcript.

11. I. T. Tichenor, Atlanta, to Joseph Squire, Helena, June 20, and July 2, 1887, Penhale transcript.

12. Joseph Squire, Helena, to I. T. Tichenor, Atlanta, December 28, 1887, Penhale transcript; I. T. Tichenor, Atlanta, to Joseph Squire, Helena, December 31, 1887, and January 2, 1888, Penhale transcript.

13. Eugene A. Smith, Tuscaloosa, to Joseph Squire, Helena, February 26, 1888, Penhale transcript; Joseph Squire, Helena, to I. T. Tichenor, Atlanta, June 4, 1888, Penhale transcript.

14. Eugene A. Smith, Tuscaloosa, to Joseph Squire, Helena, June 8 and 15, 1889, Penhale transcript; Joseph Squire, Helena, to Eugene A. Smith, Tuscaloosa, June 11, 1889, Penhale transcript.

15. Eugene A. Smith, Tuscaloosa, to Joseph Squire, Helena, October 4, 1889, and April 2 and 14, 1890, Penhale transcript; Joseph Squire, Helena, to Eugene A. Smith, Tuscaloosa, April 7 and 12, 1890, Penhale transcript.

16. Joseph Squire, Helena, to Eugene A. Smith, Tuscaloosa, June 6 and 14, and July 1, 1890, Penhale transcript. Squire sent numbered diagrams (1–36) to Julius Bien & Co., 139 Duane Street, New York. Eugene A. Smith, Tuscaloosa, to Joseph Squire, Helena, June 29, 1890, Penhale transcript; Joseph Squire, "Autobiography, 1908(?)" (photocopy of the handwritten manuscript Squire prepared for Ethel Armes in possession of author), 41.

17. Ethel Armes, *The Story of Coal and Iron in Alabama* (Birmingham, AL: The Chamber of Commerce, 1910), 152; Joseph Squire, Helena, to A. J. Reach and Co., December 28, 1888, Archives, Birmingham Public Library, Birmingham; Eugene A. Smith, Tuscaloosa, to Joseph Squire, Helena, November 9, 1888, Penhale transcript.

18. Joseph Squire, Helena, to Eugene A. Smith, Tuscaloosa, October 12, 1893, November 4 and 8, 1893, and February 20, 1896, Penhale transcript; Joseph Squire, Helena, to Truman H. Aldrich, Birmingham, September 21, 1896, Penhale transcript. Squire informed Aldrich that he had completed ninety field maps in the past four years. Eugene A. Smith, Tuscaloosa, to Joseph Squire, Helena, October 9 and 14, and November 6, 1893, Penhale transcript; Armes, *Coal and Iron,* 153.

19. Eugene A. Smith, *Geological Survey of Alabama: Report of Progress for 1875* (Montgomery: W. W. Screws, 1876), 30–32.

20. Long-wall mining is defined as "a method of working a coal seam believed to have originated in England in the 17th century. The seam is removed in one operation by means of a long working face or wall advancing in a continuous line which may be several hundred yards long." "Mine Terminology" in James H. Walker, *The Struggle and the Joy: An American Coal Town—Piper, Alabama* (Birmingham, AL: Birmingham Public Library Press, 1993), 237-242. Smith, *Report of Progress for 1875*, 30–32; *Birmingham Iron Age,* May 28, 1874, and July 22, 1875.

21. W. David Lewis, *Sloss Furnaces and the Rise of the Birmingham District: An Industrial Epic* (Tuscaloosa: University of Alabama Press, 1994), 39, 43. Lewis reports that 1875 marked a watershed because "for the first time, more pig iron was smelted in the United States with bituminous coal and coke—947,545 as against 908,046 tons—than with anthracite. By this time, only 410,990 of the national output of 2,226,581 tons of pig iron was still made with charcoal. This decline was not because charcoal iron was inferior in quality; . . . coke, however, was cheaper and superior for high-volume production." Wayne Cline, *Alabama Railroads* (Tuscaloosa: University of Alabama Press, 1997), 13. Justin Fuller, "Boom Towns and Blast Furnaces: Town Promotion in Alabama, 1885–1893," *Alabama Review* 29, no. 1 (January 1976), 37–38.

22. Smith, *Report of Progress for 1875*, 34–38.

23. Armes, *Coal and Iron,* 201, 239; S. F. H. Tarrant, *Hon. Daniel Pratt: A Biography* (Richmond, VA: Whittet & Shepperson, 1904), 56; Justin Fuller, "Henry F. DeBardeleben, Industrialist of the New South," *Alabama Review* 39, no. 1 (January 1986), 4. Prior to March 1874, the Eureka Company received mail at the "Pratt" post office, but "Ironton" referred to the rail station. This revivification resulted in the universal use of "Oxmoor" to identify the site. *Birmingham Iron Age,* March 26, 1874.

24. Armes, *Coal and Iron,* 255–256. Armes identifies other members of the governing body as A. J. Noble, treasurer; T. S. Mount, secretary; and George Gilmer, Charles T. Pollard, David Clopton, B. S. Bibb, and M. E. Pratt, directors. Formerly, 196.5 bushels of charcoal produced eight tons of iron. Under Goodrich's supervision, 123 bushels of charcoal produced eighteen tons of iron. The "Gospel" according to Goodrich was twofold: the conversion from charcoal to coke and the combination of a red and brown ore mix with coke and limestone would produce less expensive pig iron. *Birmingham Iron Age,* May 21, 1874.

25. Armes, *Coal and Iron,* 258.

26. Lewis, *Sloss Furnaces,* 64. Lewis refutes Armes's statement that coke iron was produced on February 28, 1876, basing his conclusion on a report from the *Birmingham Weekly Iron Age* that gives the March 11 date (March 16, 1876). In addition, he notes that "the account in *BWIA* indicates that James Thomas, not Goodrich, superintended the Oxmoor experiment. Armes placed Thomas at the Irondale furnace at this time." Lewis, *Sloss Furnaces,* 526n53.

27. Armes, *Coal and Iron,* 272.

28. Fuller, "DeBardeleben," 5.

29. Armes, *Coal and Iron,* 277–278.

30. Ibid., 275, 287–288; Lewis, *Sloss Furnaces,* 69.

31. Fuller, "DeBardeleben," 6; Armes, *Coal and Iron,* 295–297.

32. Armes, *Coal and Iron,* 295–297.

33. Cline, *Alabama Railroads,* 127.

34. Ibid., 129. Armes, *Coal and Iron,* 297–298. Pensacola had challenged Mobile since 1874 as the best port for Alabama coal, and many coal operators deemed it the "Gulf port of choice." *Birmingham Iron Age,* June 25, 1874, and July 22, 1875.

35. Ethel Armes identifies Ensley's "Memphis crowd" as Napoleon Hill, J. C. Nealy, W. N. Milburn, and William Fontaine. She also states that each of these capitalists "was then at feud with every mother's son of the Tennessee Coal, Iron, and Railroad Company, up in Tennessee, and deeply desirous, as was natural in Tennessee, to exterminate all rivals from the face of the globe." Armes, *Coal and Iron,* 290, 306–307. Fuller, "DeBardeleben," 8.

36. Fuller, "DeBardeleben," 11.

37. Ibid., 11–12; Armes, *Coal and Iron,* 339.

38. Lewis, *Sloss Furnaces,* 137; Justin Fuller, "History of the Tennessee Coal, Iron, and Railroad Company, 1852–1907" (PhD diss., University of North Carolina-Chapel Hill, 1966), 45.

39. Fuller, "History of TCI," 78.

40. Lewis, *Sloss Furnaces,* 155, 178; Fuller, "Boom Towns," 47.

41. Fuller, "History of TCI," 101–104.

42. Ibid., 105–106.

43. Ibid., 106–108; Armes, *Coal and Iron,* 425. Ethel Armes describes TCI as a conglomerate of twelve separate companies—the Sewanee Mining Company; the Sewanee Furnace Company; the Southern States Coal, Iron, and Land Company; the Pratt Coal and Iron Company consisting of the Alice Furnace Company, the Pratt Coal and Coke Company, and the Linn Iron Works; the DeBardeleben Coal and Iron Company; the Eureka Furnace Company; the Little Belle Furnace Company; the Henry Ellen Company; the Cahaba Coal Mining Company; and the Excelsior Mining Company.

44. John Mack Faragher et al., *Out of Many: A History of the American People,* brief 3rd ed. (Upper Saddle River, NJ: Prentice Hall, 2001), 356. According to Faragher, "Between 1866 and 1897, fourteen years of prosperity stood against seventeen of hard times. The two major depressions of 1873–79 and 1893–97 were the worst in the nation's history up to that time. Three 'minor' recessions (1866–67, 1883–85, and 1890–91) did not seem insignificant to the millions who lost their jobs." Fuller, "History of TCI," 122–123. Armes, *Coal and Iron,* 427.

45. Fuller, "History of TCI," 124–125.

46. William Bullard Allen, "Memoirs—Truman Heminway Aldrich," American Society of Civil Engineers, March 7, 1933, transcript in possession of author, 2.

47. Agreement between Truman H. Aldrich and Joseph Squire, April 20, 1895, Penhale transcript; Truman H. Aldrich Jr., Winfield, to Joseph Squire, Helena, December 20, 1898, Penhale transcript; Truman H. Aldrich, Hargrove, to Joseph Squire, Helena,

April 4, 1896, Penhale transcript; Truman H. Aldrich to Joseph Squire, Helena, January 18, and February 24, 1900, and January 22, 1906, Penhale transcript; Truman H. Aldrich Jr., Alexander City, to Joseph Squire, Helena, May 17, 1906, Penhale transcript; Joseph Squire, Helena, to Truman H. Aldrich, Winfield, December 27, 1898, Penhale transcript; Joseph Squire, Helena, to Truman H. Aldrich, January 25, 1906, Penhale transcript.

48. Allen, "Memoirs—Truman Heminway Aldrich," 3. Allen quotes Dr. Walter B. Jones, successor to Dr. Eugene A. Smith, state geologist: "[Mr. Aldrich] is considered as one of the four most eminent tertiary paleontologists of all time, along with T. A. Conrad, Isaac Lea, and Sir Charles Lyell." W. David Lewis identifies two business strategies—"developmental" and "opportunistic." "Developmental strategy was aimed at making gradual profits through careful management of resources, prudent reinvestment of income, and close attention to promoting the economic growth of areas served by the rail system. By contrast, opportunistic strategy was based on pursuing quick returns through over-rapid expansion, frequent reorganizations, stock manipulation, and other predatory tactics at the expense of sound financial management, effective maintenance of equipment and facilities, efficient service to the public, and patient development of newly entered areas." Lewis, *Sloss Furnaces,* 108. See also Maury Klein, *The Great Richmond Terminal: A Study in Businessmen and Business Strategy* (Charlottesville: University Press of Virginia, 1970), 12–13, 16–29, 288–294; Arthur M. Johnson and Barry E. Supple, *Boston Capitalists and Western Railroads: A Study in the Nineteenth-Century Railroad Investment Process* (Cambridge: Harvard University Press, 1967), 8–10, 181–191, 333–346; Fuller, "DeBardeleben," 16.

49. Lewis, *Sloss Furnaces,* 285.

Chapter 4

1. Named by Truman Aldrich for a "block" of coal estimated to weigh one "ton," "Blockton" superceded the original name of Gresham. County historian Rhoda Ellison suggests that the earlier name stemmed from some connection with the local family of Mrs. E. H. Gressham. Rhoda Coleman Ellison, *Place Names of Bibb County, Alabama* (Brierfield, AL: Cahaba Trace Commission, 1993), 75. Local historian Charles Adams challenges this theory, asserting that the town's first postmaster, Cornelius Cadle, sought to honor one of his Civil War commanders, General Walter Quinton Gresham of Indiana. As Adams points out, Gresham served as U.S. postmaster general in 1883. Charles Edward Adams, *Blocton: The History of an Alabama Coal Mining Town* (Brierfield, AL: Cahaba Trace Commission, 2001), 1–2; *Bibb (AL) Blade,* September 22, 1881, and September 17, 1884.

2. Adams, *Blocton,* 6, 9–11.

3. W. B. Allen, ed., "Sewanee Mining and Tennessee Coal, Iron, and Railroad Company Records," 533 84–148 MH3, Birmingham Public Library Archives, Birmingham, AL, 137; Adams, *Blocton,* 5, 12, 22–23, 33, 64.

4. A "mantrip" consisted of cars loaded with miners entering or leaving the mine. Rather than hauling coal ("coal trip"), these trips transported shifts of miners to and from

the coal face. The "manway" constituted a separate slope that allowed miners to walk or ride to and from their respective work stations. A cross section of Number 1 Mine depicted the following strata from bottom to top: coal (five feet), rash (six inches), coal (ten inches), slate (eighteen inches), coal (eighteen inches), slate (three feet), coal (ten inches).

5. Henry E. Colton, "Alabama Coal Mines—Blocton," *Shelby Sentinel,* July 21, 1886; Adams, *Blocton,* 13–14, 16–17, 25.

6. Patricia Sachs, "Together We Work, Together We Grow Old: Life, Work and Community in a Coal Mining Town" (PhD diss., The City University of New York, 1986), 82.

7. One resident described Number 3 Town's location as follows: "[From Blocton] the road ran east from the L&N depot and passed under the Number 2 railroad track, and past the large meat market; across the road on a bridge, then over the tracks of the E.T.V.&G. (East Tennessee, Virginia, and Georgia) Railroad." Adams, *Blocton,* 20–21. S. L. Johnson, "Brief History of Bibb County," *Centreville (AL) Press,* May 26, and June 16 and 23, 1949.

8. Allen, "Sewanee Mining and TCI," 130–139. TCI also opened mines Number 9 and 10, commencing operations in 1899.

9. Thomas Dublin and Walter Licht, *The Face of Decline: The Pennsylvania Anthracite Region in the Twentieth Century* (Ithaca, NY: Cornell University Press, 2005), 199; Thomas G. Andrews, *Killing for Coal: America's Deadliest Labor War* (Cambridge: Harvard University Press, 2008), 17, 188, 230. For textiles, see Jacquelyn Dowd Hall et al., *Like a Family: The Making of a Southern Cotton Mill World,* Fred W. Morrison Series in Southern Studies (Chapel Hill: University of North Carolina Press, 2000); and Allen Tullos, *Habits of Industry: White Culture and the Transformation of the Carolina Piedmont,* Fred W. Morrison Series in Southern Studies (Chapel Hill: University of North Carolina Press, 1989).

10. Henry A. Emfinger, *The Story of My Hometown, Aldrich, Alabama* (Montevallo, AL: privately printed, 1969), 8–14, 51. In similar fashion, Colorado Fuel and Iron built Segundo with houses "painted in different pleasing colors" to counter "monotonous uniformity." Andrews, *Killing for Coal,* 212–213.

11. Ibid., 14–15.

12. Crandall A. Shifflett, *Coal Towns: Life, Work, and Culture in Company Towns of Southern Appalachia, 1880–1960* (Knoxville: University of Tennessee Press, 1991), xiii.

13. Ibid., xviii, 48; James H. Walker, *The Struggle and the Joy: An American Coal Town—Piper, Alabama* (Birmingham, AL: Birmingham Public Library Press, 1993), xii; Marie Butler, *Margaret, Alabama . . . and Now There's Gold: A Brief History of a Unique Community* (Birmingham, AL: privately printed, 1989), ix.

14. Shifflett, *Coal Towns,* 27–30.

15. Ibid., xiii–xvi, 4–11; Wayne Flynt, *Poor but Proud: Alabama's Poor Whites* (Tuscaloosa: University of Alabama Press, 1989), 124.

16. "Other named seams are Atkins, Wadsworth, Gould, Polecat, Maylene, Lovelady, Wooten, Air Shaft, Yessick, Little Pittsburgh, Brock and Earhart." State of Alabama, *Annual Report of Coal Mines* (Birmingham, AL: Dispatch Printing Co., 1928), 21–23.

17. Shifflett, *Coal Towns,* 5–6; Walker, *Struggle and Joy,* 124. Flynt, *Poor but Proud,* 127–128.

18. Sachs, "Together We Work," 84–85.

19. Adams, *Blocton,* 17–18; Walker, *Struggle and Joy,* 61–63.

20. Sachs, "Together We Work," 86; Daniel Letwin, *The Challenge of Interracial Unionism: Alabama Coal Miners, 1878–1921* (Chapel Hill: University of North Carolina Press, 1997), 22; Shifflett, *Coal Towns,* 87–89.

21. Flynt, *Poor but Proud,* 128; Shifflett, *Coal Towns,* 85–86. See also H. E. Mills, *Coal Mining in Alabama: A Series of Information Letters* (Montgomery: The Alabama Mining Institute, 1936).

22. Jack Roland Bergstresser Sr., "Raw Material Constraints and Technological Options in the Mines and Furnaces of the Birmingham District: 1876–1930" (PhD diss., Auburn University, 1993), 56, 61, 75, 85, 89; Alabama, *Annual Report of Coal Mines* (1921), 74–75. The *American Heritage Dictionary* defines "specific gravity" as the ratio of the mass of a solid . . . to the mass of an equal volume of distilled water at 4 degrees centigrade.

23. U.S. Department of the Interior, *A Medical Survey of the Bituminous Coal Industry: Report of the Coal Mines Administration* (Washington, DC: Government Printing Office, 1947), 4; Shifflett, *Coal Towns,* xiii, 41. Flynt, *Poor but Proud,* 116.

24. Shifflett, *Coal Towns,* 34.

25. This and subsequent data concerning the number of employees, type and amount of coal, pay scales, days worked, hoisting, and ventilation are taken from the *Annual Report of Coal Mines,* 1913–1929, 1934–1938, FY 1944–1948, and from State of Alabama, Department of Industrial Relations, Division of Safety and Inspection, *Annual Statistical Report* (Wetumpka, AL: Wetumpka Printing Co., FY 1951–1991).

26. Some place names are taken from Ellison, *Place Names of Bibb County,* 29–141 passim.

27. Henry M. McKiven Jr., *Iron and Steel: Class, Race, and Community in Birmingham, Alabama, 1875–1920* (Chapel Hill: University of North Carolina Press, 1995), 61. Anthracite operations in Pennsylvania also used standard British work plans, methods, and technology. Anthony F. C. Wallace, *St. Clair: A Nineteenth-Century Coal Town's Experience with a Disaster-Prone Industry* (New York: Alfred A. Knopf, 1987), 171.

28. Walker, *Struggle and Joy,* 11–13.

29. Ibid., 13–19.

30. Ibid., 2, 5, 10, 21, 25, 119. Cecil O. Sewell Sr., "As I Remember Piper, Alabama: 1928–1941," interview by James H. Walker, in Walker, *Struggle and Joy,* 42.

31. Walker, *Struggle and Joy,* 9–10, 20–21; Walter Gardner, "By-Gone Days," *Centreville (AL) Press,* April 17, 1980.

32. In addition to the Eureka and Paramount operations, the Helena Mine tapped the Black Shale Seam, operating from 1917 to 1921. These miners hauled coal to the Atlanta, Birmingham & Atlantic Railroad by wagon.

33. Information concerning Straven and the Sicard family is taken from Mary Sicard Cobb Russo, "History of Sicard and Mas Families, as of July 1991," transcript in possession of author, and from Dr. L. C. Parnell of Birmingham, interview by author, February 28, 1998, Birmingham.

34. Parnell, 1998; *Heritage of Shelby County, Alabama* (Clanton, AL: Heritage Pub-

lishing Consultants, Inc., 1999), 165–166. Henry Brantley of the nearby Lacey community purchased many of the Straven structures, using the lumber and other materials to construct houses in present-day Brantleyville.

35. Butler, *Margaret,* 7, 23, 41. Margaret was named for DeBardeleben's daughter-in-law, wife of Charles F. DeBardeleben, and "Acmar" derives from the first letters of the AF&IC's other mines, *Ac*ton + *Mar*garet. See Mattie Lou Teague Crow, *History of St. Clair County (Alabama)* (Huntsville, AL: The Strode Publishers, 1973), 133; Marilyn Davis Barefield, *A History of Hoover, Alabama, and Its People* (Birmingham, AL: The Hoover Historical Society, 1992), 43–46. Local legends state that the Reverend William Acton and his son, Zephaniah William Henry Acton, mined coal near the Cahaba River in the 1840s. Contrary to this author's focus on William Phineas Browne, Zephaniah's obituary claims that he and his father mined the first coal in Jefferson and Shelby Counties.

36. Butler, *Margaret,* 3; Ronald L. Lewis, *Black Coal Miners in America: Race, Class, and Community Conflict, 1780–1980* (Lexington: University Press of Kentucky, 1987), 72; Edwin L. Brown and Colin J. Davis, eds., *It Is Union and Liberty: Alabama Coal Miners and the UMW* (Tuscaloosa: University of Alabama Press, 1999), 51; Brian Kelly, *Race, Class, and Power in the Alabama Coalfields, 1908–21* (Urbana: University of Illinois Press, 2001), 151.

37. Barefield, *A History of Hoover,* 44; Butler, *Margaret,* 9.

38. Alabama, *Annual Report of Coal Mines* (1923), 102–103.

39. *Blocton (AL) Enterprise,* July 16, 1908; Adams, *Blocton,* 127–129.

40. Johnson, *Centreville (AL) Press,* June 16, and July 7, 1949; Adams, *Blocton,* 20–22, 84. Similarly, Wallace reports that St. Clair's operations constituted a "patchwork" of ethnic groups—English, Welsh, and Scottish. Wallace, *St. Clair,* 171.

41. *Blocton (AL) Courier,* December 4, 1891. In 1890, local businesses included: West Blocton—furniture and undertaking, general merchandise, clothing, dry goods, watches and jewelry, groceries, drugs and medicines, fruits and produce, butcher, blacksmith and woodwork, cobbler, newspaper, hotel, livery stable, saw mill and lumber; Blocton—general merchandise, soda fountain, fruits, watches, books and stationery, newspapers and magazines, cobbler, barbershop, hotel, fish and meats, photographer, blacksmith; Smith Hill—dry goods, groceries, millinery, general merchandise, bakery, meat. Adams, *Blocton,* 39–42, 192.

42. Shifflett, *Coal Towns,* xv, 60–65; Flynt, *Poor but Proud,* 121; William Warren Rogers, Robert David Ward, Leah Rawls Atkins, and Wayne Flynt, *Alabama: The History of a Deep South State* (Tuscaloosa: University of Alabama Press, 1994), 283; Adams, *Blocton,* 83.

43. *Blocton (AL) Courier,* December 4, 1891; Adams, *Blocton,* 42, 47, 161.

44. Rhoda Coleman Ellison, *Bibb County, Alabama: The First Hundred Years, 1818–1918* (Tuscaloosa: University of Alabama Press, 1984), 170–172. Ellison reports that an Italian boy served as interpreter for the census worker in 1890. Adams, *Blocton,* 71; Patti B. Watson, "A Special Place in Time," interview by James H. Walker, in Walker, *Struggle and Joy,* 37–38; Harry Fullman, interview by James H. Walker, in Walker, *Struggle and Joy,* 135–136; Marie Samsal Fancher, interview by James H. Walker (October 1990), in Walker, *Struggle and Joy,* 139–140.

45. Ellison, *Bibb County, Alabama,* 172, 288. "'Dago' from the Spanish *Diego* ('James') was a term that had long been used contemptuously of Spaniards, Portuguese, and Italians." Rhoda Coleman Ellison, "Little Italy in Rural Alabama," *Alabama Heritage* 2 (Fall 1986), 34–45; Johnson, *Centreville (AL) Press,* June 16, 1949; Adams, *Blocton,* 20–22, 47–48.

46. Shifflett, *Coal Towns,* 51, 79; Ellison, quoted in Adams, *Blocton,* 131.

47. *Blocton (AL) Enterprise,* August 5, 1915; Adams, *Blocton,* 120, 131, 133–134, 136, 138, 140, 142. The third Liberty Loan Drive in April 1918 exceeded the two-week goal of $6,800 by collecting $23,850. Later, TCI sold an additional $12,000 in bonds, and the fourth Liberty Loan Drive netted $75,000.

48. Adams, *Blocton,* 138; Ellison, *Bibb County, Alabama,* 174.

Chapter 5

1. For a comprehensive account of the Banner Mine explosion, see Robert David Ward and William Warren Rogers, *Convicts, Coal, and the Banner Mine Tragedy* (Tuscaloosa: University of Alabama Press, 1987); Ronald L. Lewis, *Black Coal Miners in America: Race, Class, and Community Conflict, 1780–1980* (Lexington: University Press of Kentucky, 1987), 31.

2. Charles Edward Adams, *Blocton: The History of an Alabama Coal Mining Town* (Brierfield, AL: Cahaba Trace Commission, 2001), 124–127.

3. Alex Lichtenstein asserts that convict leasing conflicted with and competed against free labor. Alex Lichtenstein, *Twice the Work of Free Labor: The Political Economy of Convict Labor in the New South* (London: Verso, 1996), 15. See also Douglas A. Blackmon, *Slavery by Another Name: The Re-Enslavement of Black People in America from the Civil War to World War II* (New York: Doubleday, 2008).

4. Matthew J. Mancini, *One Dies, Get Another: Convict Leasing in the American South, 1866–1928* (Columbia: University of South Carolina Press, 1996), 1–3, 20, 30–31, 38; Lichtenstein, *Twice the Work,* xvii–xix, 3–5, 19, 38–43, 69–70, 103.

5. Elizabeth Boner Clark, "The Abolition of the Convict Lease System in Alabama, 1913–1928" (master's thesis, University of Alabama, 1949), ii, 1–2; William Franklin Drummond, "Utilization of Convict Labor in the South" (master's thesis, University of Alabama, 1933), 54.

6. Clark, "Abolition of the Convict Lease System," 11–12; Mancini, *One Dies, Get Another,* 99–101. Some records spell McMillan with an "E," that is, "McMillen."

7. Clark, "Abolition of the Convict Lease System," 14; Drummond, "Utilization of Convict Labor," 18–19, 37–38; Douglas A. Blackmon, "From Alabama's Past, Capitalism Teamed with Racism to Create Cruel Partnership," *Wall Street Journal,* July 16, 2001, 4; Lichtenstein, *Twice the Work,* 4, 80–81, 188.

8. Milfred C. Fierce, *Slavery Revisited: Blacks and the Southern Convict Lease System, 1865–1933* (New York: Africana Studies Research Center, Brooklyn College, City University of New York, 1994), x, 2, 15, 77, 252–253.

9. Ibid., 88, 90–91; Albert B. Moore, *History of Alabama and Her People* (New York:

The American Historical Society, 1927), 1:979, cited in Fierce, *Slavery Revisited,* 112–113; Mary Church Terrell, "Peonage in the United States: The Convict Lease System and the Chain Gangs," *The Nineteenth Century* 62 (1907), 308, cited in Fierce, *Slavery Revisited,* 231–232; W. E. B. DuBois, "The Spawn of Slavery: The Convict Lease System in the South," *Missionary Review of the World* 24 (October 1901), 744–745, cited in Fierce, *Slavery Revisited,* 240.

10. Blackmon, "From Alabama's Past," 4–5; Mary Ellen Curtin, *Black Prisoners and Their World, Alabama, 1865–1900* (Charlottesville: University Press of Virginia, 2000), 2–3, 207–210. Curtin notes that "after a massive political campaign leading to changes in Alabama's constitution, the fee system in Jefferson County was finally abolished in 1919." Curtin's research centers around correspondence filed by the Alabama Department of Corrections between 1881 and 1897. Letters exchanged among Chief Inspector Reginald Heber Dawson and his two subordinates, Dr. A. T. Henley and W. D. Lee, "contain candid accounts of daily prison life, politics, and contractors." Curtin's study complements this author's research in the Board of Inspectors annual, biennial, and quadrennial reports. Alabama Board of Inspectors of Convicts, Administrative Correspondence, 1881–1897, SG 18151–18155, Special Collections, Alabama Department of Archives and History, Montgomery, Alabama, hereafter cited as ADAH.

11. Alabama Board of Inspectors of Convicts, *Annual Report of the Inspector of the Alabama Penitentiary, 1874,* SG 13058–13060, ADAH.

12. *Annual Report, 1875,* SG 13058–13060, ADAH.

13. *Annual Report, 1876,* SG 13058–13060, 3, ADAH. *Annual Report, 1877,* SG 13058–13060, 4, ADAH. Lichtenstein notes that convict labor contributed to industrialization initially, but that the system later stagnated economic development. Lichtenstein, *Twice the Work,* 107.

14. Alabama Board of Inspectors of Convicts, *Biennial Report, 1880,* SG 13058–13060, ADAH; Alabama Department of Corrections and Institutions, Administrative Correspondence, 1909–1947, SG 17601–17604, 17606, 20624, ADAH; Alabama Board of Inspectors of Convicts, *Rules and Regulations for the Government of the Convict System of Alabama* (Montgomery, AL: Barrett & Co., 1886), 21. These *Rules and Regulations* established daily quotas for "convicts working in coal mines where coal is four feet thick or upwards" in which the miners must cut and load the coal. Further clarification revealed that "convicts loading coal and not cutting it, when coal is four feet thick or upwards, shall be required to load not exceeding the following tasks per day: First class men, Eight tons of coal. Second class men, Six tons of coal. Third class men, Four tons of coal. Fourth class men, Two tons of coal." Some reports indicate that convict miners produced more slack than lump coal. Consequently, operators tended to coke most of their output. Lichtenstein, *Twice the Work,* 89, 105–106.

15. Ezekiel Archey to R. H. Dawson, January 18, 1884, cited in Curtin, *Black Prisoners,* 20–21. An examination of the *Administrative Correspondence, 1881–1897* files used by Curtin revealed that many of the letter book copies had absorbed some moisture and that the ink had smeared. Furthermore, much of the correspondence was filed in haphazard, that is, nonchronological order. This author located enough letters to substantiate the validity

of Curtin's citations, but the lack of organization rendered the collection nonconducive to efficient historical research. Consequently, notations will cite Curtin's book as a reliable secondary source rather than focusing on the more obscure primary sources.

16. John T. Milner to R. H. Dawson, June 10, 1885, cited in Curtin, *Black Prisoners,* 69–71.

17. *Huntsville Gazette,* September 3, 1881, reprinted from the *Jackson Republican,* cited in Curtin, *Black Prisoners,* 70–71.

18. Samuel L. Webb and Margaret E. Armbrester, *Alabama Governors: A Political History of the State* (Tuscaloosa: University of Alabama Press, 2001), 107–108; Clark, "Abolition of the Convict Lease System," 15.

19. J. B. Gaston, *Annual Address before the State Medical Association in Mobile upon the Sanitary Condition of Our Prisons* (1882), SG 13058–13060, ADAH; John T. Milner response to J. B. Gaston, SG 13058–13060, ADAH; *Huntsville Gazette,* December 16, 1882, cited in Curtin, *Black Prisoners,* 76–77.

20. Curtin, *Black Prisoners,* 22, 83.

21. Clark, "Abolition of the Convict Lease System," 23; *Biennial Report, 1884,* 21, SG 13058–13060, ADAH; Curtin, *Black Prisoners,* 263–264.

22. Curtin, *Black Prisoners,* 84. Curtin also notes (p. 82–83) that "critics and subsequent historians have emphasized Bankhead's dishonest relations with mining interests. Journalist Robert McKee, editor of the *Selma Argus,* alleged that Bankhead headed a 'penitentiary ring'—a corrupt alliance between Democratic party insiders and industrialists. Bankhead parlayed a two-year stint as warden of the state penitentiary, with a salary of $2,000 a year, into a wealthy political dynasty. In 1888, three years after he left the position of warden, he received a payoff of over $1,200 from [the Tennessee Coal, Iron, and Railroad Company] so as to ensure that company would continue to receive the lion's share of prison labor. . . . It seems clear that Bankhead's reforms were merely a means of getting more prisoners into coal mines and staving off criticism about the lease." Alabama Board of Inspectors of Convicts, *Quarterly Report, Spring 1886,* 65, SG 13058–13060, ADAH.

23. *Biennial Report, 1886,* 13–15, SG 13058–13060, ADAH.

24. Curtin, *Black Prisoners,* 23, 82, 94. A state law passed in 1885 stipulated that each prisoner should have his own time card. *Biennial Report, 1886,* cited in Curtin, *Black Prisoners,* 91.

25. Curtin, *Black Prisoners,* 5, 22–23, 73–74.

26. *Biennial Report, 1888,* iv, SG 13058–13060, ADAH. Also see Curtin, *Black Prisoners,* 100, 106, 111–112, 130, 138; Webb and Armbrester, *Alabama Governors,* 115–116; *Biennial Report, 1900,* SG 13058–13060, ADAH; Lichtenstein, *Twice the Work,* 91, 104.

27. Curtin, *Black Prisoners,* 5, 37, 112, 141–143.

28. Alabama General Assembly, *An Act to create a new convict-lease system for the State of Alabama, and to provide for the government, discipline and maintenance of all convicts in the State of Alabama,* February 14, 1893, 21, SG 13058–13060, ADAH; Webb and Armbrester, *Alabama Governors,* 119, 125; Alabama General Assembly, *An Act to regulate the management of state and county convicts by General Assembly of Alabama,* February 18, 1895, 864, 876, SG 13058–13060, ADAH.

29. Curtin, *Black Prisoners,* 84, 131, 147–148, 165. In 1890, the convict-lease system generated $113,000 for Alabama's treasury—about 6 percent of the state budget. Alabama Board of Inspectors of Convicts, *Biennial Report, 1888–1906; Quadrennial Report, 1910,* SG 13058–13060, ADAH. The following data illustrate the rising importance of the convict-lease system as a source of state revenue:

1888—$103,332.97
1890—$209,579.44
1892—$225,648.84
1894—$240,736.78
1896—$260,919.14
1898—$325,196.10
1900—$361,284.98
1902—$302,870.42
1904—$767,381.05 (approximately 1,800 convicts)
1906—$1.3 million (2,059 convicts)
1910—$2.72 million (1,905 convicts)

In general, convict leasing returned profits nearly four times operating expenses. Lichtenstein, *Twice the Work,* 18–19.

30. Webb and Armbrester, *Alabama Governors,* 143–144, 153–156.

31. *Rules and Regulations for the Government of the Convict System of Alabama, Adopted by The Board of Inspectors of Convicts, March 3, 1886* (Montgomery, AL: Barrett & Co., 1886). Curtin, *Black Prisoners,* 166. Webb and Armbrester, *Alabama Governors,* 161.

32. Alabama Board of Inspectors of Convicts, *Quadrennial Report, 1910 and 1914,* SG 13058–13060, ADAH. Henry A. Emfinger, *The Story of My Hometown, Aldrich, Alabama* (Montevallo, AL: privately printed, 1969), 48. Convict labor ended for the Red Feather Coal Company at Lucile on January 1, 1918. Resident prisoners were transferred to the Bessemer Coal, Iron, and Land Company interest at Belle Ellen. That enterprise filed its last report regarding convict hire invoices with the Alabama state auditor in January 1924. See Alabama State Auditor, "Convict Hire Invoices (1875–)," SG 5248, 20658, 20667, 20668, 20673, 20674, 20679, 20680, 20685, 20686, ADAH; Adams, *Blocton,* 127.

33. *Quadrennial Report, 1910, 1914, 1918, 1922,* SG 13058–13060, ADAH. Convicts worked 310 days per year on average as compared to 170 days for free laborers. See Melvyn Stokes and Rick Halpern, eds., *Race and Class in the American South Since 1890* (Oxford: Berg Publishers, 1994), 16.

34. George Washington Cable, *The Silent South, Together with the Freedman's Case in Equity and the Convict Lease System* (New York, 1889), 126, 128, cited in Ronald L. Lewis, *Black Coal Miners,* 14; Clark, "Abolition of the Convict Lease System," 34–36.

35. As noted earlier, additional compensation for county officials frequently extended time served, credited at 30 cents per day. Ronald L. Lewis, *Black Coal Miners,* 32–35. See also Lichtenstein, *Twice the Work,* 13–14, 17–18.

36. Gaston, *Annual Address,* ADAH. Also, Curtin states that, "compared to the death rates in other prisons throughout the nation, Alabama came in first, or last, depending on perspective [64 per 1,000]. The Missouri penitentiary's rate was 16 per 1,000; Ohio's was 9 per 1,000; Maryland's was 27 per 1,000. The only state coming close to Alabama was Mississippi, whose death rate was 40 per 1,000 in 1894." Curtin, *Black Prisoners,* 155. Clark, "Abolition of the Convict Lease System," 26.

37. Clark, "Abolition of the Convict Lease System," 54–57, 70–71.

38. Ibid., 72, 88; Alabama, *Annual Report of Coal Mines* (1924–1927).

39. Mancini, *One Dies, Get Another,* 115–116, 228–230; Clark, "Abolition of the Convict Lease System," 105–106, 114. Lichtenstein states: "The chain gangs which built the roads of the twentieth-century South became an enduring symbol of southern backwardness, brutality, and racism; in fact, they were the embodiment of the Progressive ideals of southern modernization, penal reform, and racial moderation. In this duality the southern chain gang replicated the most significant feature of the convict lease system it had superseded." Lichtenstein, *Twice the Work,* 16.

40. Mancini, *One Dies, Get Another,* 115–116, 228–230; Isadore Shapiro, *Address to the American Prison Association,* New Orleans, 1917, cited in Clark, "Abolition of the Convict Lease System," 29.

Chapter 6

1. Henry A. Emfinger, *The Story of My Hometown, Aldrich, Alabama* (Montevallo, AL: privately printed, 1969), 13, 15, 43–45, 48. Truman Aldrich partnered with Peter B. Thomas in 1905, and Albert Latady joined the pair in 1908. At this point, the owners changed the name from the Montevallo Coal and Transportation Company to the Montevallo Mining Company. In 1912, the trio sold the mines to W. S. Lovell and Henry Bedham who introduced convict leasing to the community. Bankrupt by 1922, this duo left Aldrich, and the mines passed into receivership. In 1925, Durias A. Thomas, son of former owner Peter B. Thomas, purchased the Aldrich mines for $45,000 cash and $25,000 payable to the State of Alabama.

2. James H. Walker, *The Struggle and the Joy: An American Coal Town—Piper, Alabama* (Birmingham, AL: Birmingham Public Library Press, 1993),57, 111; Jess Shepard, interview by James H. Walker (December 1990), in Walker, *Struggle and Joy,* 115; Cecil O. Sewell Sr., "As I Remember Piper, Alabama: 1928–1941," interview by James H. Walker, in Walker, *Struggle and Joy,* 50, 54.

3. Marie Butler, *Margaret, Alabama . . . and Now There's Gold: A Brief History of a Unique Community* (Birmingham, AL: privately printed, 1989), vii, 2.

4. Ibid., 7–9.

5. Marilyn Davis Barefield, *A History of Hoover, Alabama, and Its People* (Birmingham, AL: The Hoover Historical Society, 1992), 46–47; Butler, *Margaret,* 10–12; David Alan Corbin, *Life, Work, and Rebellion in the Coal Fields: The Southern West Virginia Miners, 1880–1922* (Urbana: University of Illinois Press, 1981), 123–124.

6. Butler, *Margaret,* 10, 14, 37. "Smokey" Turner won the turkey by loading twenty-six carloads. Corbin contends that these types of company institutions and services, "regardless of their expense and benefit," equated to company control. Corbin, *Life, Work, and Rebellion,* 134. Colorado Fuel and Iron built paternalistic company towns "to inculcate subservience and loyalty in an increasingly diverse, persistently militant workforce." Thomas G. Andrews, *Killing for Coal: America's Deadliest Labor War* (Cambridge: Harvard University Press, 2008), 197.

7. Colorado Fuel and Iron constructed "ideal communities" with the expectation that miners would exchange "solidarity and militancy for company loyalty and industrial harmony." Andrews, *Killing for Coal,* 203–204.

8. See Carole Marks, *Farewell—We're Good and Gone: The Great Black Migration* (Bloomington: Indiana University Press, 1989).

9. Crandall A. Shifflett, *Coal Towns: Life, Work, and Culture in Company Towns of Southern Appalachia, 1880–1960* (Knoxville: University of Tennessee Press, 1991), xiii.

10. Ibid., xv, 1, 48.

11. Ibid., xv, 33, 66–67.

12. Corbin, *Life, Work, and Rebellion,* 78, 118.

13. Stuart D. Brandes, *American Welfare Capitalism, 1880–1940* (Chicago: University of Chicago Press, 1976), ix, 10–11, 14.

14. Ibid., 5–6, 8.

15. Ibid., 25–27.

16. Ibid., 32–33; Marlene Hunt Rikard, "An Experiment in Welfare Capitalism: The Health Care Services of the Tennessee Coal, Iron and Railroad Company" (PhD diss., University of Alabama, 1983), 7.

17. Brandes, *American Welfare Capitalism,* 28–29, 31, 38.

18. Ibid., 33, 35–36.

19. Walker, *Struggle and Joy.* Andrews states that Colorado miners wanted "safety, fellowship, a higher quality of life, autonomy, dignity, and basic freedoms." Andrews, *Killing for Coal,* 231.

20. John Hays Hammond, *Report of the United States Coal Commission* (Washington, DC: Government Printing Office, 1925); Shifflett, *Coal Towns,* 146–148; Jingsheng Dong, "The Social and Working Lives of Alabama Coal Miners and Their Families, 1893–1925," (master's thesis, Auburn University, 1988), 51.

21. S. L. Johnson, "Brief History of Bibb County," *Centreville (AL) Press,* 9 June 1949; Shifflett, *Coal Towns,* 176; Walker, *Struggle and Joy,* 69, 91.

22. Emfinger, *Aldrich,* 14, 57; Patti B. Watson, "A Special Place in Time," interview by James H. Walker, in Walker, *Struggle and Joy,* 38; Walker, *Struggle and Joy,* 125–126; Butler, *Margaret,* 26.

23. Shifflett, *Coal Towns,* 189; Walker, *Struggle and Joy,* 72. Eventually the LCCC consolidated its resources by appointing one manager for both stores. Brandes, *American Welfare Capitalism,* 45.

24. Shifflett, *Coal Towns,* 178, 188–189; Wayne Flynt, *Poor but Proud: Alabama's Poor*

Whites (Tuscaloosa: University of Alabama Press, 1989), 118; Sewell, "As I Remember Piper," in Walker, *Struggle and Joy,* 48; Walker, *Struggle and Joy,* 69–70, 72–73.

25. Dong, "Social and Working Lives," 19–20, 25–26, 40; Flynt, *Poor but Proud,* 119–120. For a comprehensive index of various forms of scrip, see Donald O. Edkins, *Edkins' Catalogue of United States Coal Company Store Scrip (Tokens)* (New Kensington, PA: Catalogue Committee of the National Scrip Collectors Association, 1977); Shifflett, *Coal Towns,* 183.

26. Flynt, *Poor but Proud,* 119; Shifflett, *Coal Towns,* 179–180.

27. Shifflett, *Coal Towns,* 182; Butler, *Margaret,* 39; Flynt, *Poor but Proud,* 118.

28. Brandes, *American Welfare Capitalism,* 83, 85, 95, 98; Charles Edward Adams, *Blocton: The History of an Alabama Coal Mining Town* (Brierfield, AL: Cahaba Trace Commission, 2001), 24–25.

29. Brandes, *American Welfare Capitalism,* 107.

30. Ibid., 87.

31. Ibid., 98, 101; Rikard, "Experiment in Welfare Capitalism," 75, 122, 133, 161.

32. Shifflett, *Coal Towns,* 55–56; Rikard, "Experiment in Welfare Capitalism," 139–145, 166, 189.

33. Brandes, *American Welfare Capitalism,* 113, 118. "Partly because of its relief aspect, social work was one of the few aspects of welfare capitalism maintained during economic depressions." Rikard, "Experiment in Welfare Capitalism," 108, 245.

34. *Blocton (AL) Enterprise,* January 10, 1910, cited in Adams, *Blocton,* 119–120.

35. Sewell, "As I Remember Piper," in Walker, *Struggle and Joy,* 46. For an additional 50 cents per month, Piper residents could obtain a private hospital room.

36. Butler, *Margaret,* 42.

37. Brian Kelly, *Race, Class, and Power in the Alabama Coalfields, 1908–21* (Urbana: University of Illinois Press, 2001), 85–86; Daniel Letwin, *The Challenge of Interracial Unionism: Alabama Coal Miners, 1878–1921* (Chapel Hill: University of North Carolina Press, 1997), 102; Philip Taft, *Organizing Dixie: Alabama Workers in the Industrial Era* (Westport, CT: Greenwood Press, 1981), 85–86.

38. Ronald L. Lewis, *Black Coal Miners in America: Race, Class, and Community Conflict, 1780–1980* (Lexington: University Press of Kentucky, 1987), 74–75. Lewis also lists mining conditions as reported by federal investigators and recorded in the *United Mine Workers Journal:*

> Fourteen-hour workdays; quotas which had to be met before leaving work; no lunch break; a system known as 'share cropping,' whereby every third car a miner loaded was credited to the contractor; armed guards who also served as labor contractors; the three-thousand-pound ton, and the miners bore the loss from washing; no checkweighman; an elaborate system of fines for dirty coal; no pay for dead work; deductions from meager paychecks for such things as rent for preachers, company stores which kept men in debt; churches which preached nonunionism along with religion, a strict rule requiring every housewife to raise a garden and can no less

than a hundred quarts per year; the requirement of special permission for visitors entering the barbed-wire camps and special permits for miners who wanted to leave; segregated, company-run unions; and an elaborate system of espionage.

39. Brandes, *American Welfare Capitalism,* 43–44, 47.

40. Ibid., 40–41; Flynt, *Poor but Proud,* 116–118; Adams, *Blocton,* 20–21, 24–25; Johnson, *Centreville (AL) Press,* June 16, 1949; Emfinger, *Aldrich,* 52; Butler, *Margaret,* 24; Walker, *Struggle and Joy,* 65–66. Long-time residents gave identity to their houses, that is, certain places might be known as the "Smith House" or the "Jones House" even though the company remained the lawful owner.

41. Johnson, *Centreville (AL) Press,* March 29, 1951; Adams, *Blocton,* 24–25, 33; Emfinger, *Aldrich,* 51; Butler, *Margaret,* 24; Sewell, "As I Remember Piper," in Walker, *Struggle and Joy,* 55. In his study of Colorado miners, Andrews notes that, "through exterior landscaping and interior design, welfare capitalists sought to turn company houses into sanctuaries where mining families could lead orderly middle-class lives, protected from the temptations of street, saloon, and workscape." Andrews, *Killing for Coal,* 225.

42. Eventually, the West Blocton town council passed an ordinance prohibiting cows, pigs, and horses from the downtown area. Adams, *Blocton,* 189–190; Sewell, "As I Remember Piper," in Walker, *Struggle and Joy,* 55–56; Walker, *Struggle and Joy,* 66–67; Walter Gardner, "By-Gone Days," *Centreville (AL) Press,* August 7, 1980. Some houses included a barn or shed for larger animals. Families without a milk cow received deliveries of milk, buttermilk, and butter from local dairies.

43. Johnson, *Centreville (AL) Press,* July 5, 1951; Adams, *Blocton,* 18–19. Walker, *Struggle and Joy,* 66–67. Before electricity, ice companies serviced iceboxes located on the back porch or in the kitchen. A "weight card" located on the front porch could be turned to inform the deliveryman of the amount of ice desired. Families oriented the card so that the bottom number designated the appropriate weight or block size, and they left payment on top of the icebox.

44. Flynt, *Poor but Proud,* 120; Shifflett, *Coal Towns,* 190–196; Liston Pope, *Millhands and Preachers: A Study of Gastonia* (New Haven, CT: Yale University Press, 1942), 21–23.

45. *Blocton (AL) Courier,* December 4, 1891; Jesse Cole Persinger, "I Have Kept the Faith" (n.p.: privately printed, 1988), 38; *Blocton (AL) Enterprise,* August 4, 1910; Adams, *Blocton,* 42, 51–58, 160–161. The Episcopal congregation leased land from TCI in 1900 ($1 for 99 years) and consecrated a church building in 1906. In 1923, the Baptists moved to West Blocton, and the Catholics initiated Lenten services in three languages—Latin, Italian, and English.

46. Willie Jo Parker, West Blocton, to Charles Adams, University, Summer 1983, transcript in possession of Charles E. Adams, Tuscaloosa, AL; Adams, *Blocton,* 58–61.

47. Emfinger, *Aldrich,* 14, 52, 60–63; Butler, *Margaret,* 26.

48. Brandes, *American Welfare Capitalism,* 53–58. In West Virginia, school systems represented a "moral" and "educational" agent for sociocultural change. Corbin, *Life, Work, and Rebellion,* 127.

49. Brandes, *American Welfare Capitalism,* 35, 59–60; Adams, *Blocton,* 42, 195. The

1890 census reported that 2,841 of 7,966 coal miners (35.7 percent) were illiterate. By 1920, this figure had declined to 14 percent of 26,000 miners.

50. Brandes, *American Welfare Capitalism,* 60–65. Colorado Fuel and Iron used a company newspaper, the *Camp and Plant,* to enhance its public image. Andrews, *Killing for Coal,* 209–211.

51. Rikard, "Experiment in Welfare Capitalism," 247; George M. Neal, interview by Wade Taylor, November 3, 1982, transcript in possession of Charles E. Adams, Tuscaloosa, AL; Adams, *Blocton,* 102, 107–108, 166.

52. *Centreville County (AL) Press,* December 9, 1897; Parker to Adams, Summer 1983; *Blocton (AL) Enterprise,* March 11, 1909; Adams, *Blocton,* 43–47, 103–104. Blocton's Number 1 School included students from Piper, Belle Ellen, Hargrove, and Braehead. Also, the Catholic parish school opened in 1901.

53. Adams, *Blocton,* 95–97, 153–155, 162. In 1903, West Blocton received a net profit of $1,200 from the dispensary system (gross sales equaled $4,100) and an additional $6,827.25 from the Bibb County Board of Education. *Centreville (AL) Press,* February 7, and August 22, 1907.

54. Walker, *Struggle and Joy,* 99–100. The "Joe Lee" transported students from Marvel, Seymour, Coleanor, Piper, and Belle Ellen. Dr. Stinson was instrumental in providing the lighting for the basketball and tennis courts. Fire destroyed the school building in 1940; the elementary school was rebuilt, but buses transported high school students to West Blocton.

55. Emfinger, *Aldrich,* 51–52; Butler, *Margaret,* 26–29, 33.

56. Shifflett, *Coal Towns,* 149; Adams, *Blocton,* 211.

57. Shifflett, *Coal Towns,* 162–164. Eventually, baseball's popularity declined due to four factors: (1) unionism (1930s)—companies no longer sponsored teams, bought uniforms, organized leagues; (2) World War II (1940s)—military service changed miners' overall outlook; (3) closing of company towns—baseball teams lost their "identity, support, spectators"; (4) automobile and television—provided alternative forms of recreation. *Blocton (AL) Courier,* June 17, 1892; Johnson, *Centreville (AL) Press,* July 5, 1951; Adams, *Blocton,* 33–37.

58. Adams, *Blocton,* 39, 110, 133–134, 149, 163, 169. Blue laws prohibited motion pictures on Sunday. *Blocton (AL) Enterprise,* June 20, 1907.

59. Adams, *Blocton,* 185, 195–196. Corbin argues that movies and theater productions became instruments "through which the operators attempted to inculcate new values and ideas." Corbin, *Life, Work, and Rebellion,* 126.

60. *Blocton (AL) Courier,* December 4, 1891; *Blocton (AL) Enterprise,* November 3, 1906; Adams, *Blocton,* 33, 37, 115–117, 159–160, 194.

61. *Centreville (AL) Enterprise,* January 2, 1890; *Blocton (AL) Courier,* December 4, 1891; *Centreville (AL) Press,* July 18, 1907; Adams, *Blocton,* 42, 48–50, 117–119, 153, 211. Adams also mentions the Benevolent and Protective Order of Elks, Independent Order of Beavers, Order of the Owls, Yellowhammers Nest, Knights Templar, and Woodmen of the World.

62. Johnson, *Centreville (AL) Press,* December 2, 1948; Adams, *Blocton,* 119, 153, 211;

Butler, *Margaret,* 41–42; Sewell, "As I Remember Piper," in Walker, *Struggle and Joy,* 43; Walter Gardner Jr., "Christmas in Piper and Coleanor," interview by James H. Walker, in Walker, *Struggle and Joy,* 133–134.

63. David Brody, "The Rise and Decline of Welfare Capitalism," in John Braeman et al., *Change and Continuity in Twentieth-Century America: The Twenties* (Columbus: Ohio State University Press, 1968), 160–161, 171; Brandes, *American Welfare Capitalism,* 136, 140–141.

Chapter 7

1. Herschel H. Day, "A Brief History of the Andrew J. Day Family of Bibb County," interview by Mary Fuzzell, transcript in possession of author.

2. Philip Taft, *Organizing Dixie: Alabama Workers in the Industrial Era* (Westport, CT: Greenwood Press, 1981), 74; Marlene Hunt Rikard, "An Experiment in Welfare Capitalism: The Health Care Services of the Tennessee Coal, Iron and Railroad Company" (PhD diss., University of Alabama, 1983), 307–308.

3. Stuart D. Brandes, *American Welfare Capitalism, 1880–1940* (Chicago: University of Chicago Press, 1976), 142; Rikard, "Experiment in Welfare Capitalism," 352; Richard Edwards, *Contested Terrain: The Transformation of the Workplace in the Twentieth Century* (New York: Basic Books, Inc., 1979), 95–97. According to Andrews, paternalism "fanned the flames of unrest in Colorado." Thomas G. Andrews, *Killing for Coal: America's Deadliest Labor War* (Cambridge: Harvard University Press, 2008), 199. In West Virginia, the company town represented the heart of an authoritarian system that imposed social control on mining families. Complete and ruthless rule by owners often drove miners to strike to maintain their dignity and freedom. David Alan Corbin, ed., *The West Virginia Mine Wars: An Anthology* (Martinsburg, WV: Appalachian Editions, 1990), 1.

4. Taft, *Organizing Dixie,* 88.

5. Major Smith's even-handed approach was an aberration. As Andrews reports, militia officers usually sided with coal company executives, not with coal field migrants or union leaders. Andrews, *Killing for Coal,* 257. Piper historian James H. Walker cites B. L. Thompson, a worker at Piper No. 1 and father of six, as an example. Thompson worked eleven days in February 1934 and earned $37.50. The company then deducted the following amounts: company store charges, $23.00; house rent, $7.20; lamp rent, $0.90; doctor's services, $1.20; school fees, $0.85; flowers for funeral, $0.20. Balance due was $4.15. Cecil O. Sewell Sr., "As I Remember Piper, Alabama: 1928–1941," interview by James H. Walker, in James H. Walker, *The Struggle and the Joy: An American Coal Town—Piper, Alabama* (Birmingham, AL: Birmingham Public Library Press, 1993), 52; Walker, *Struggle and Joy,* 29–34.

6. Marie Butler, *Margaret, Alabama . . . and Now There's Gold: A Brief History of a Unique Community* (Birmingham, AL: privately printed, 1989), 3, 16–18; Taft, *Organizing Dixie,* 91–93.

7. Taft, *Organizing Dixie,* 87; Shaunna L. Scott, *Two Sides to Everything: The Cul-*

tural Construction of Class Consciousness in Harlan County, Kentucky (New York: State University of New York Press, 1995), 221. Strikes also occurred in 1933 at the Aldrich, Peerless, and Little Gem mines.

8. Crandall A. Shifflett, *Coal Towns: Life, Work, and Culture in Company Towns of Southern Appalachia, 1880–1960* (Knoxville: University of Tennessee Press, 1991), 118.

9. Daniel Letwin, *The Challenge of Interracial Unionism: Alabama Coal Miners, 1878–1921* (Chapel Hill: University of North Carolina Press, 1997), 35; William Warren Rogers, Robert David Ward, Leah Rawls Atkins, and Wayne Flynt, *Alabama: The History of a Deep South State* (Tuscaloosa: University of Alabama Press, 1994), 293.

10. Herbert G. Gutman, "Black Coal Miners and the Greenback-Labor Party in Redeemer, Alabama, 1878–79: The Letters of Warren D. Kelley, Willis Johnson Thomas, 'Dawson,' and Others," *Labor History* 10, no. 3 (Summer 1969), 506–510.

11. Ibid., 515, 533–534; Wayne Flynt, *Poor but Proud: Alabama's Poor Whites* (Tuscaloosa: University of Alabama Press, 1989), 138.

12. The history of unionism in the Cahaba coal field closely parallels that of organized labor throughout Alabama. Several reliable monographs focus on respective strike periods and organizing efforts. Therefore, this author will cite these secondary sources in an attempt to synthesize the development of labor unions within the state. Edwin L. Brown and Colin J. Davis, eds., *It Is Union and Liberty: Alabama Coal Miners and the UMW* (Tuscaloosa: University of Alabama Press, 1999), 14–15.

13. Letwin, *Challenge of Interracial Unionism,* 55–56, 68–73.

14. The Labor Party of Alabama attempted unsuccessfully to combine the interests of the Knights of Labor, the Wheel, the Farmers' Alliance, the Tailors' Union, the Typographical Union, and the Carpenters' Union. Robert David Ward and William Warren Rogers, *Labor Revolt in Alabama: The Great Strike of 1894* (University: University of Alabama Press, 1965), 38; Letwin, *Challenge of Interracial Unionism,* 86, 96.

15. Letwin, *Challenge of Interracial Unionism,* 45–48; Flynt, *Poor but Proud,* 138–139.

16. Letwin, *Challenge of Interracial Unionism,* 90; Maier B. Fox, *United We Stand: The United Mine Workers of America, 1890–1990* (n.p.: UMWA, 1990), 27, 110–111; Paul B. Worthman, "Black Workers and Labor Unions in Birmingham, Alabama, 1897–1904," *Labor History* 10, no. 3 (Summer 1969), 381, 390. Black miners habitually served as vice presidents for district and local organizations, in three of eight positions on the district executive board, and on all functional and ceremonial committees. Also, Worthman notes that black convention delegates did not fill traditional roles defined by subordination and deference. Rather, they "actively participated in committee discussions and floor debates." See also John H. M. Laslett, *Labor and the Left: A Study of Socialist and Radical Influences in the American Labor Movement, 1881–1924* (New York: Basic Books, 1970), 192–240.

17. Ward and Rogers, *Labor Revolt,* 31; Letwin, *Challenge of Interracial Unionism,* 91.

18. Letwin, *Challenge of Interracial Unionism,* 91; Ward and Rogers, *Labor Revolt,* 34.

19. Brown and Davis, *It Is Union and Liberty,* 17–18.

20. Letwin, *Challenge of Interracial Unionism,* 99; Taft, *Organizing Dixie,* 21; Ward and Rogers, *Labor Revolt,* 51, 56. Similarly, miners in southern West Virginia initially re-

mained outside the UMWA, "not out of apathy or cowardice, but because the issues and goals of the miners' union did not represent their wants and needs." David Alan Corbin, *Life, Work, and Rebellion in the Coal Fields: The Southern West Virginia Miners, 1880–1922* (Urbana: University of Illinois Press, 1981), 26.

21. Ward and Rogers, *Labor Revolt,* 9, 49–63. Ward and Rogers contend that "Alabama's organized labor movement was strongly Populist, so that the often cited lack of cooperation between union men and Populists did not hold true in Alabama."

22. Letwin, *Challenge of Interracial Unionism,* 101–102. UMWA historian Maier B. Fox states that "some of the state's operators had invited the confrontation as a means of introducing convicts to the mines and preventing the restoration of the old wage scale." Fox, *United We Stand,* 45; Alex Lichtenstein, *Twice the Work of Free Labor: The Political Economy of Convict Labor in the New South* (London: Verso, 1996), 102.

23. Ward and Rogers, *Labor Revolt,* 45, 77. Jones had commended coal miners in November for their nonviolent resistance to the convict-lease system, and he "showed understanding and even sympathy for the miners' opposition to convict labor." Fox, *United We Stand,* 46.

24. Some Blocton mines closed, but Numbers 1, 2, 3, and 5 continued operations. Output from these four mines constituted half the 1893 production. Rhoda Coleman Ellison, *Bibb County, Alabama: The First Hundred Years, 1818–1918* (Tuscaloosa: University of Alabama Press, 1984), 176; Ward and Rogers, *Labor Revolt,* 95–115. Blocton remained a stronghold for union activity, even inviting socialist Eugene Debs to speak on two separate occasions in 1896. Charles Edward Adams, *Blocton: The History of an Alabama Coal Mining Town* (Brierfield, AL: Cahaba Trace Commission, 2001), 70.

25. Letwin, *Challenge of Interracial Unionism,* 109. Andrews notes that Colorado Fuel and Iron successfully opposed the Strike of 1894 by using spatial and social control techniques to reduce labor costs and to prevent labor organization. In turn, this action prompted the implementation of social welfare work, welfare capitalism, and industrial paternalism. Ultimately, Colorado's Ten Days' War of 1914 constituted a violent reaction to the spatial and social order imposed by Colorado Fuel and Iron in 1894. Andrews, *Killing for Coal,* 196–198, 214.

26. Ward and Rogers, *Labor Revolt,* 136; Brown and Davis, *It Is Union and Liberty,* 23; Letwin, *Challenge of Interracial Unionism,* 116.

27. Letwin, *Challenge of Interracial Unionism,* 113; Brown and Davis, *It Is Union and Liberty,* 24–25. Taft records that "no unified group opposition to the right of the union arose prior to 1904." Taft, *Organizing Dixie,* 26; Worthman, "Black Workers," 388, 399.

28. Fox, *United We Stand,* 136–137; Letwin, *Challenge of Interracial Unionism,* 142–143.

29. *Centreville (AL) Press,* November 17 and 24, 1904; *Birmingham Age-Herald,* November 18, 1904; Adams, *Blocton,* 90–91; Rikard, "Experiment in Welfare Capitalism," 25.

30. Letwin, *Challenge of Interracial Unionism,* 145; Fox, *United We Stand,* 136–137.

31. Brown and Davis, *It Is Union and Liberty,* 3; Brian Kelly, *Race, Class, and Power in the Alabama Coalfields, 1908–21* (Urbana: University of Illinois Press, 2001), 19; Flynt, *Poor but Proud,* 140–141.

32. Adams, *Blocton,* 99–100; Ellison, *Bibb County,* 176–177; Flynt, *Poor but Proud,* 140–141.

33. Brown and Davis, *It Is Union and Liberty,* 34. Daniel Letwin provides the following metaphor: "Race-baiting never found a more combustible setting: the coarsening racial temper of the region provided the atmosphere; the conspicuous role of white women in an interracial strike, the tinder; the 'social equality' charge, the match." Letwin, *Challenge of Interracial Unionism,* 154; Rogers et al., *Alabama,* 370; Fox, *United We Stand,* 136–137; Kelly, *Race, Class, and Power,* 27. See also Richard A. Straw, "The Collapse of Biracial Unionism: The Alabama Coal Strike of 1908," *Alabama Historical Quarterly* 37, no. 2 (Summer 1975), 92–114; and Richard A. Straw, "Soldiers and Miners in a Strike Zone: Birmingham, 1908," *Alabama Review* 38, no. 4 (October 1985), 289–308.

34. Letwin, *Challenge of Interracial Unionism,* 160–161; Flynt, *Poor but Proud,* 143.

35. Flynt, *Poor but Proud,* 141–143; Brown and Davis, *It Is Union and Liberty,* 36–37, 54, 71. Taft notes that union membership had declined to twenty-eight by 1914. Taft, *Organizing Dixie,* 46.

36. Brown and Davis, *It Is Union and Liberty,* 7, 39–45.

37. Kelly, *Race, Class, and Power,* 105–106, 135–141; Letwin, *Challenge of Interracial Unionism,* 165–168, 176.

38. Letwin, *Challenge of Interracial Unionism,* 159; Kelly, *Race, Class, and Power,* 134, 156.

39. Fox, *United We Stand,* 182–183. "In September [1918] the War Department exempted all coal miners from military conscription." Corbin, *Life, Work, and Rebellion,* 178.

40. Kelly, *Race, Class, and Power,* 164–168; Letwin, *Challenge of Interracial Unionism,* 174–184; Taft, *Organizing Dixie,* 51.

41. Adams, *Blocton,* 139, 151.

42. Kelly, *Race, Class, and Power,* 169; Letwin, *Challenge of Interracial Unionism,* 184; Adams, *Blocton,* 151. Numerous violent mass strikes in 1919 sought to preserve wartime gains and to promote the ideals of freedom, industrial democracy, social equality, and political rights. Corbin, *Life, Work, and Rebellion,* 176; Corbin, *West Virginia Mine Wars,* ii.

43. Kelly, *Race, Class, and Power,* 162, 173.

44. Mamie Trott Day, Piper, to Annie Trott Gibbons, Birmingham, October 15, 1920, transcript in possession of author.

45. Kelly, *Race, Class, and Power,* 177–179, 184.

46. Ibid., 197; Letwin, *Challenge of Interracial Unionism,* 185–187. See also Richard A. Straw, "The United Mine Workers of America and the 1920 Coal Strike in Alabama," *Alabama Review* 28, no. 2 (April 1975), 104–128.

47. Brown and Davis, *It Is Union and Liberty,* 62; Flynt, *Poor but Proud,* 135–137; Kelly, *Race, Class, and Power,* 14. Similarly, Corbin views strikes in West Virginia as "extensions of a world view, a working-class culture." Corbin, *Life, Work, and Rebellion,* xviii.

48. Charles Adams relates that at Blocton's Number 9 Black Diamond mine, if someone "accidentally spilled the water from his lunch bucket, the rest of the crew would pour out theirs in sympathy and walk off the job." Adams, *Blocton,* 243; Brown and Davis, *It Is*

Union and Liberty, 70, 78; Letwin, *Challenge of Interracial Unionism,* 189; Rogers et al., *Alabama,* 481–484; Shifflett, *Coal Towns,* 139.

49. Rikard, "Experiment in Welfare Capitalism," 1; Shifflett, *Coal Towns,* xvi, 155.

Chapter 8

1. Lizzie Frost of Calera, AL, interview by author, October 2, 2000, Calera.

2. Wayne Flynt, *Poor but Proud: Alabama's Poor Whites* (Tuscaloosa: University of Alabama Press, 1989), 118, 126, 133–134; James H. Walker, *The Struggle and the Joy: An American Coal Town—Piper, Alabama* (Birmingham, AL: Birmingham Public Library Press, 1993), 123–124; Charles Edward Adams, *Blocton: The History of an Alabama Coal Mining Town* (Brierfield, AL: Cahaba Trace Commission, 2001), 203; Rhoda Coleman Ellison, *Bibb County, Alabama: The First Hundred Years, 1818–1918* (Tuscaloosa: University of Alabama Press, 1984), 174. Charles Adams reports that miners could dig four or five tons per day and, at 40 cents per ton, earn about $10 per week. Wayne Flynt provides wage rates that illustrate changes over time:

1892—sorting slate/coal (age 12)	11–12 cents/day
1893—mule driver (age 13)	18 cents/day
1896—boss driver of teams (age 16)	22 cents/day
1914—mining coal (avg. 5–10 tons/day)	29 cents/ton
1920—mining coal (12-hour day)	$1.85/day
1925—mining coal (age 15)	$2.00/day
1903—foreman	$75.00/month
1920—foreman	$900.00/month

3. Crandall A. Shifflett, *Coal Towns: Life, Work, and Culture in Company Towns of Southern Appalachia, 1880–1960* (Knoxville: University of Tennessee Press, 1991), 212.

4. Burton R. Morley, *Characteristics of the Labor Market in Alabama Related to the Administration of Unemployment Compensation* (University: Bureau of Business Research, University of Alabama, 1937), 29–30, 55, 63; Shifflett, *Coal Towns,* 203.

5. Walker, *Struggle and Joy,* xii, 23–24, 46–47; Cecil O. Sewell Sr., "As I Remember Piper, Alabama: 1928–1941," interview by James H. Walker, in Walker, *Struggle and Joy,* 55–56.

6. U.S. Department of the Interior, *A Medical Survey of the Bituminous Coal Industry: Report of the Coal Mines Administration* (Washington, DC: Government Printing Office, 1947), 149; Flynt, *Poor but Proud,* 130–131.

7. Safety data for subsequent years may be found in Charles S. Perry and Christian Ritter, *Dying to Dig Coal: Fatalities in Deep and Surface Coal Mining in Appalachian States, 1930–1978* (Lexington: Department of Sociology, University of Kentucky, 1981); Department of Interior, *Medical Survey,* 151; Marilyn Davis Barefield, *A History of Hoover, Alabama, and Its People* (Birmingham, AL: The Hoover Historical Society, 1992), 46–47; Marie Butler, *Margaret, Alabama . . . and Now There's Gold: A Brief History of a Unique*

Community (Birmingham, AL: privately printed, 1989), 10, 48; Flynt, *Poor but Proud,* 131–132.

8. Harry Fullman, interview by James H. Walker, in Walker, *Struggle and Joy,* 137–138.

9. Walker, *Struggle and Joy,* 24–27; Department of Interior, *Medical Survey,* 286.

10. Walker, *Struggle and Joy,* 22; Sewell, "As I Remember Piper," in Walker, *Struggle and Joy,* 46–47.

11. Duane Lockard, *Coal: A Memoir and Critique* (Charlottesville: University Press of Virginia, 1998), 40–41; Patricia Sachs, "Together We Work, Together We Grow Old: Life, Work and Community in a Coal Mining Town" (PhD diss., The City University of New York, 1986), 90. See also Keith Dix, *What's a Coal Miner to Do? The Mechanization of Coal Mining* (Pittsburgh: University of Pittsburgh Press, 1988).

12. Perry and Ritter, *Dying to Dig Coal,* 10.

13. Lockard, *Coal,* 41–44.

14. Ibid., 44–47.

15. Shifflett, *Coal Towns,* 210.

16. Ibid., 203–206.

17. Ibid., 9–10, 206. Adams, *Blocton,* 122–123, 134–135, 160–161, 164. Shifflett asserts that the automobile ultimately allowed third-generation miners to abandon the company towns altogether. In essence, "mobility, to some degree, was always a source of the miner's freedom and independence."

18. Adams, *Blocton,* 170–190. See also, Charles Edward Adams, "The Great West Blocton Town Fire of 1927," *Alabama Heritage* 49 (Summer 1998), 34–40.

19. Adams, *Blocton,* 347–349, 356–359, 365.

20. Henry A. Emfinger, *The Story of My Hometown, Aldrich, Alabama* (Montevallo, AL: privately printed, 1969), 52; *Centreville (AL) Press,* March 9, 1950; Adams, *Blocton,* 244–245, 247, 251–253. Aldrich historian Henry A. Emfinger records 1947 as Boothton's last year of operation.

21. Walker, *Struggle and Joy,* 61–63; Walter Gardner, "By-Gone Days," *Centreville (AL) Press,* April 17, 1980, 12.

22. Sewell, "As I Remember Piper," in Walker, *Struggle and Joy,* 56–57; Walker, *Struggle and Joy,* 156–159, 174.

23. Morley, *Characteristics of Labor Market,* 15–16; *Heritage of Shelby County, Alabama* (Clanton, AL: Heritage Publishing Consultants, Inc., 1999), 96–97. Woodrow Hand recounts a similar story about "Johnnie Fence, Truck Miner" in James Seay Brown Jr., ed., *Up Before Daylight: Life Histories from the Alabama Writers' Project, 1938–1939* (Tuscaloosa: University of Alabama Press, 1982), 100–105.

24. Virgil Rice of Vestavia Hills, AL, interview by author, March 20, 2000, Hoover, Alabama.

25. State of Alabama, *Annual Report of Coal Mines* (Birmingham, AL: Dispatch Printing Co., 1936–1938, FY 1944–1948); State of Alabama, Department of Industrial Relations, Division of Safety and Inspection, *Annual Statistical Report* (Wetumpka, AL: Wetumpka Printing Co., FY 1951–1991). Bibb County—two to thirteen truck mines (1936–1964), one to four underground mines (1965–1968), one to nine strip mines (1965–1991). Jefferson

County—eleven to seventeen truck mines (1936–1938), one to three strip mines (1971–1988). St. Clair County—one to seven truck mines (1945–1959), one underground mine (1978), one to five strip mines (1969–1989). Shelby County—eleven to twenty-eight truck mines (1936–1964), one to fourteen underground mines (1965–1972), one to five strip mines (1967–1989).

26. Emfinger, *Aldrich,* 8, 49, 51; Butler, *Margaret,* 51–52; Barefield, *History of Hoover,* 47. Ironically, the Harbert Corporation converted the Acton mining area into the upscale Riverchase real estate development.

27. Ellison, *Bibb County,* 184; Adams, *Blocton,* 192, 197–200, 203–204.

28. Adams, *Blocton,* 204–205, 211.

29. *Centreville (AL) Press,* March 1, 1938; Adams, *Blocton,* 212–213, 219.

30. *Centreville (AL) Press,* November 28, 1946; Adams, *Blocton,* 220–222, 234, 238–240.

31. Adams, *Blocton,* 246–247, 253, 255–259.

32. Ibid., 266, 278, 286–288, 300.

33. Alabama, *Annual Report of Coal Mines* (1913–1929, 1934–1938, FY 1944–1948); Alabama, *Annual Statistical Report* (FY 1951–1991); Ed Flynn. "Report of Inspector of Alabama Coal Mines," Birmingham, AL: Freret & Grant, 1910.

SELECT BIBLIOGRAPHY

Adams, Charles Edward. *Blocton: The History of an Alabama Coal Mining Town.* Brierfield, AL: Cahaba Trace Commission, 2001.

———. "The Great West Blocton Town Fire of 1927." *Alabama Heritage* 49 (Summer 1998), 34–40.

Adams, Sean Patrick. *Old Dominion, Industrial Commonwealth: Coal, Politics, and Economy in Antebellum America*. Baltimore: The Johns Hopkins University Press, 2004.

Alabama Blast Furnaces. Woodward, AL: Woodward Iron Company, 1940.

Alabama Board of Inspectors of Convicts. Administrative Correspondence, 1881–1897. Files SG 18151–18155. Alabama Department of Archives and History, Montgomery, AL.

———. Quarterly, Annual, Biennial, and Quadrennial Reports of the Inspector of the Alabama Penitentiary, 1845–1922. Files SG 13058–13060. Alabama Department of Archives and History, Montgomery, AL.

———. *Rules and Regulations for the Government of the Convict System of Alabama.* Montgomery, AL: Barrett & Co., 1886.

Alabama Department of Corrections and Institutions. Administrative Correspondence, 1909–1947. Files SG 17601–17604, 17606, 20624. Alabama Department of Archives and History, Montgomery, AL.

Alabama General Assembly. *An Act To Create a New Convict-Lease System for the State of Alabama, and To Provide for the Government, Discipline and Maintenance of All Convicts in the State of Alabama,* February 14, 1893. Files SG 13058–13060. Alabama Department of Archives and History, Montgomery, AL.

———. *An Act To Regulate the Management of State and County Convicts by General Assembly of Alabama,* February 18, 1895. Files SG 13058–13060. Alabama Department of Archives and History, Montgomery, AL.

———. *Acts of the [45th] Session of 1872–73 of the General Assembly of Alabama Held in the City of Montgomery, Commencing November 18th, 1872.* Montgomery, AL: Arthur Bingham, 1873.

———. *Acts of the [52d] General Assembly of Alabama, Passed at the Session of 1882–83, Held in The City of Montgomery, Commencing 2d Tuesday in November 1882.* Montgomery, AL: W. D. Brown & Co., 1883.

Alabama State Auditor. "Convict Hire Invoices (1875–)." Files SG 5248, 20658, 20667,

20668, 20673, 20674, 20679, 20680, 20685, 20686. Alabama Department of Archives and History, Montgomery, AL.

Aldrich, T. H. "Historical Account of Coal Mining Operations in Alabama Since 1853." In *Geological Survey of Alabama: Report of Progress for 1875,* edited by Eugene A. Smith, 28–45. Montgomery, AL: W. W. Screws, 1876.

Allen, W. B., ed. "Sewanee Mining and Tennessee Coal, Iron, and Railroad Company Records." 533 84–148 MH3. Birmingham Public Library Archives, Birmingham, AL.

Allen, William Bullard. "Memoirs—Truman Heminway Aldrich." American Society of Civil Engineers, March 7, 1933. Transcript in the possession of the author.

Andrews, Thomas G. *Killing for Coal: America's Deadliest Labor War.* Cambridge: Harvard University Press, 2008.

Armes, Ethel. *The Story of Coal and Iron in Alabama.* Birmingham, AL: The Chamber of Commerce, 1910.

Barefield, Marilyn Davis. *A History of Hoover, Alabama, and Its People.* Birmingham, AL: The Hoover Historical Society, 1992.

Bennett, James R. *Tannehill and the Growth of the Alabama Iron Industry.* McCalla: Alabama Historic Ironworks Commission, 1999.

Bergstresser, Jack Roland, Sr. "Raw Material Constraints and Technological Options in the Mines and Furnaces of the Birmingham District: 1876–1930." PhD diss., Auburn University, 1993.

Blackmon, Douglas A. "From Alabama's Past, Capitalism Teamed with Racism to Create Cruel Partnership." *Wall Street Journal,* July 16, 2001.

———. *Slavery by Another Name: The Re-Enslavement of Black People in America from the Civil War to World War II.* New York: Doubleday, 2008.

Brandes, Stuart D. *American Welfare Capitalism, 1880–1940.* Chicago: University of Chicago Press, 1976.

Brody, David. "The Rise and Decline of Welfare Capitalism." In *Change and Continuity in Twentieth-Century America: The Twenties,* edited by John Braeman, Robert H. Bremner, David Brody, 147–179. Columbus: Ohio State University Press, 1968.

Brown, Edwin L., and Colin J. Davis, eds. *It Is Union and Liberty: Alabama Coal Miners and the UMW.* Tuscaloosa: University of Alabama Press, 1999.

Brown, James Seay, Jr., ed. *Up Before Daylight: Life Histories from the Alabama Writers' Project, 1938–1939.* Tuscaloosa: University of Alabama Press, 1982.

Browne, William Phineas. "Browne Family Papers." File LPR 132. Alabama Department of Archives and History, Montgomery, AL.

Butler, Marie. *Margaret, Alabama . . . and Now There's Gold: A Brief History of a Unique Community.* Birmingham, AL: privately printed, 1989.

Caudill, Harry M. *A Darkness at Dawn: Appalachian Kentucky and the Future.* Lexington: University Press of Kentucky, 1976.

Clark, Elizabeth Boner. "The Abolition of the Convict Lease System in Alabama, 1913–1928." Master's thesis, University of Alabama, 1949.

Clemmons, Vicky, and David Daniel, on behalf of the Centreville Historic Preservation Commission. *Bibb County.* Image of America Series. Charleston, SC: Arcadia Publishing, 2008.

Cline, Wayne. *Alabama Railroads.* Tuscaloosa: University of Alabama Press, 1997.

"'Coal Field of Tuscaloosa, Alabama'—being an extract of a letter to Prof. Silliman, from Charles Lyell, Esq., dated Mobile, Alabama, Feb. 19th, 1846." *The American Journal of Science* 1 (May 1846), 371–376.

Colton, Henry E. "Alabama Coal Mines—Blocton." *Shelby Sentinel,* July 21, 1886.

Corbin, David Alan. *Life, Work, and Rebellion in the Coal Fields: The Southern West Virginia Miners, 1880–1922.* Urbana: University of Illinois Press, 1981.

———, ed. *The West Virginia Mine Wars: An Anthology.* Martinsburg, WV: Appalachian Editions, 1990.

Crow, Mattie Lou Teague. *History of St. Clair County (Alabama).* Huntsville, AL: The Strode Publishers, 1973.

Curtin, Mary Ellen. *Black Prisoners and Their World, Alabama, 1865–1900.* Charlottesville: University Press of Virginia, 2000.

Day, Herschel H. "A Brief History of the Andrew J. Day Family of Bibb County." Interview by Mary Fuzzell. Transcript in the possession of the author.

Day, Mamie Trott, Piper, to Annie Trott Gibbons, Birmingham. Transcript in the possession of the author.

Dean, Lewis S., ed. *Michael Tuomey's Reports and Letters on the Geology of Alabama, 1847–1856.* Information Series 77. Tuscaloosa: Geological Survey of Alabama, 1995.

Dix, Keith. *What's a Coal Miner to Do? The Mechanization of Coal Mining.* Pittsburgh: University of Pittsburgh Press, 1988.

Dong, Jingsheng. "The Social and Working Lives of Alabama Coal Miners and Their Families, 1893–1925." Master's thesis, Auburn University, 1988.

Drummond, William Franklin. "Utilization of Convict Labor in the South." Master's thesis, University of Alabama, 1933.

Dublin, Thomas, and Walter Licht. *The Face of Decline: The Pennsylvania Anthracite Region in the Twentieth Century.* Ithaca, NY: Cornell University Press, 2005.

Edkins, Donald O. *Edkins' Catalogue of United States Coal Company Store Scrip (Tokens).* New Kensington, PA: The Catalogue Committee of The National Scrip Collectors Association, 1977.

Edwards, Richard. *Contested Terrain: The Transformation of the Workplace in the Twentieth Century.* New York: Basic Books, 1979.

Ellison, Rhoda Coleman. *Bibb County, Alabama: The First Hundred Years, 1818–1918.* Tuscaloosa: University of Alabama Press, 1984.

———. "Little Italy in Rural Alabama." *Alabama Heritage* 2 (Fall 1986), 34–47.

———. *Place Names of Bibb County, Alabama.* Brierfield, AL: Cahaba Trace Commission, 1993.

Emfinger, Henry A. *The Story of My Hometown, Aldrich, Alabama.* Montevallo, AL: privately printed, 1969.

Evans, Curtis J. *The Conquest of Labor: Daniel Pratt and Southern Industrialization.* Baton Rouge: Louisiana State University Press, 2001.

Everse, Martin. "Sally's Story." *Shelby County Historical Quarterly* 14–16 (April 1988), 28–40.

Faragher, John Mack, Mari Jo Buhle, Daniel H. Czitrom, Susan H. Armitage. *Out of*

Many: A History of the American People, Brief 3rd ed. Upper Saddle River, NJ: Prentice Hall, 2001.

Fierce, Milfred C. *Slavery Revisited: Blacks and the Southern Convict Lease System, 1865–1933.* New York: Africana Studies Research Center, Brooklyn College, City University of New York, 1994.

Flynn, Ed. "Report of Inspector of Alabama Coal Mines." Birmingham, AL: Freret & Grant, 1910.

Flynt, Wayne. *Poor but Proud: Alabama's Poor Whites.* Tuscaloosa: University of Alabama Press, 1989.

Fox, Maier B. *United We Stand: The United Mine Workers of America, 1890–1990.* [Washington, DC]: UMWA, 1990.

Frost, Lizzie of Calera, AL. Interview by author, October 2, 2000, Calera.

Fuller, Justin. "Boom Towns and Blast Furnaces: Town Promotion in Alabama, 1885–1893." *Alabama Review* 29, no. 1 (January 1976), 37–48.

———. "Henry F. DeBardeleben, Industrialist of the New South." *Alabama Review* 39, no. 1 (January 1986), 3–16.

———. "History of the Tennessee Coal, Iron, and Railroad Company, 1852–1907." PhD diss., University of North Carolina—Chapel Hill, 1966.

Gadd, Ben. *Bankhead: The Twenty Year Town.* Banff, Alberta, Canada: The Friends of Banff National Park, 1989.

Gardner, Walter, Jr. "By-Gone Days." *Centreville (AL) Press,* March 20–August 7, 1980.

Gerstner, Patsy. *Henry Darwin Rogers, 1808–1866: American Geologist.* Tuscaloosa: University of Alabama Press, 1994.

Gutman, Herbert G. "Black Coal Miners and the Greenback-Labor Party in Redeemer, Alabama, 1878–1879: The Letters of Warren D. Kelley, Willis Johnson Thomas, 'Dawson,' and Others." *Labor History* 10, no. 3 (Summer 1969), 506–535.

Hall, Jacquelyn Dowd, James Leloudis, Robert Korstad, Mary Murphy, Lu Ann Jones, Christopher B. Daly. *Like a Family: The Making of a Southern Cotton Mill World.* Fred W. Morrison Series in Southern Studies. Chapel Hill: University of North Carolina Press, 2000.

Hammond, John Hays. *Report of the United States Coal Commission.* Washington, DC: Government Printing Office, 1925.

Harrison, Tommie. *Good Morning Yesterday.* Alabaster, AL: privately printed, 1977.

———. *Wilton: Sketches of the Past.* Alabaster, AL: privately printed, 1982.

Heritage of Bibb County, Alabama. Clanton, AL: Heritage Publishing Consultants, Inc., 1998.

Heritage of Shelby County, Alabama. Clanton, AL: Heritage Publishing Consultants, Inc., 1999.

Hultquist, Clark, and Carey Heatherly. *Montevallo.* Images of America Series. Charleston, SC: Arcadia Publishing, 2011.

Johnson, Arthur M., and Barry E. Supple. *Boston Capitalists and Western Railroads: A Study in the Nineteenth-Century Railroad Investment Process.* Cambridge: Harvard University Press, 1967.

Johnson, Kimball. "Isaac Taylor Tichenor: A Biography." ThD thesis, Southern Baptist Theological Seminary, 1955.

Johnson, S. L. "Brief History of Bibb County," *Centreville (AL) Press,* December 2, 1948–July 5, 1951.

Kelly, Brian. *Race, Class, and Power in the Alabama Coalfields, 1908–21.* Urbana: University of Illinois Press, 2001.

Klein, Maury. *The Great Richmond Terminal: A Study in Businessmen and Business Strategy.* Charlottesville: University Press of Virginia, 1970.

Knapp, Virginia Estella. "William Phineas Browne, A Yankee Business Man of the South." Master's thesis, University of Texas, 1948.

Laslett, John H. M. *Labor and the Left: A Study of Socialist and Radical Influences in the American Labor Movement, 1881–1924.* New York: Basic Books, 1970.

Letwin, Daniel. *The Challenge of Interracial Unionism: Alabama Coal Miners, 1878–1921.* Chapel Hill: University of North Carolina Press, 1997.

Lewis, Ronald L. *Black Coal Miners in America: Race, Class, and Community Conflict, 1780–1980.* Lexington: University Press of Kentucky, 1987.

———. *Coal, Iron, and Slaves: Industrial Slavery in Maryland and Virginia, 1715–1865.* Westport, CT: Greenwood Press, 1979.

Lewis, W. David. *Sloss Furnaces and the Rise of the Birmingham District: An Industrial Epic.* Tuscaloosa: University of Alabama Press, 1994.

Lichtenstein, Alex. *Twice the Work of Free Labor: The Political Economy of Convict Labor in the New South.* London: Verso, 1996.

Lockard, Duane. *Coal: A Memoir and Critique.* Charlottesville: University Press of Virginia, 1998.

Lyell, Sir Charles. *A Second Visit to the United States of North America.* New York: Harper & Brothers, 1849.

Mancini, Matthew J. *One Dies, Get Another: Convict Leasing in the American South, 1866–1928.* Columbia: University of South Carolina Press, 1996.

Marks, Carole. *Farewell—We're Good and Gone: The Great Black Migration.* Bloomington: Indiana University Press, 1989.

McKiven, Henry M., Jr. *Iron and Steel: Class, Race, and Community in Birmingham, Alabama, 1875–1920.* Chapel Hill: University of North Carolina Press, 1995.

Meroney, Eloise. *Montevallo: The First One Hundred Years.* Montevallo, AL: Times Printing Company, 1977.

Millbrooke, Anne. "The Geological Society of Pennsylvania, 1832–1836, Part 1, Founding the Society." *Pennsylvania Geology* 7 (1976): 7–11.

———. "The Geological Society of Pennsylvania, 1832–1836, Part 2, Promoting a State Survey." *Pennsylvania Geology* 8 (1977): 12–16.

———. "State Geological Surveys of the Nineteenth Century." PhD diss., University of Pennsylvania, 1981.

Mills, H. E. *Coal Mining in Alabama: A Series of Information Letters.* Montgomery: The Alabama Mining Institute, 1936.

Milner, John T. *Report of the Chief Engineer to the President and Board of Directors of the South and North Alabama Railroad Co., on the 26th of November, 1859.* Montgomery, AL: Advertiser Steam Printing House, 1859.

———, Montgomery, to William B. Gilmer, Montgomery, March 17, 1864. Transcript in the possession of the author.

Morley, Burton R. *Characteristics of the Labor Market in Alabama Related to the Administration of Unemployment Compensation.* University: Bureau of Business Research, University of Alabama, 1937.

Neal, George M. Interview by Wade Taylor, November 3, 1982. Transcript in the possession of Charles E. Adams, Tuscaloosa, AL.

Owen, Thomas M., and Marie Bankhead Owen. *History of Alabama and Dictionary of Alabama Biography.* 4 vols. Chicago: S. J. Clarke Publishing Company, 1921.

Parker, Willie Jo, West Blocton, to Charles Adams, University, Summer 1983. Transcript in the possession of Charles E. Adams, Tuscaloosa, AL.

Parnell, L. C., of Birmingham. Interview by author, February 28, 1998, Birmingham, AL.

Pashin, Jack C., Richard E. Carroll, Robert L. Barnett, and Mirza A. Beg. *Geology and Coal Resources of the Cahaba Coal Field.* Bulletin 163. Tuscaloosa: Geological Survey of Alabama, 1995.

Penhale, Ken, and Martin Everse. *Helena, Alabama.* Images of America Series. Charleston, SC: Arcadia Publishing, 1998.

Perry, Charles S., and Christian Ritter. *Dying To Dig Coal: Fatalities in Deep and Surface Coal Mining in Appalachian States, 1930–1978.* Lexington: Department of Sociology, University of Kentucky, 1981.

Persinger, Jesse Cole. "I Have Kept the Faith." N.p.: privately printed, 1988.

Pope, Liston. *Millhands and Preachers: A Study of Gastonia.* New Haven, CT: Yale University Press, 1942.

Rice, Virgil, of Vestavia Hills, AL. Interview by author, March 20, 2000, Hoover, AL.

Rikard, Marlene Hunt. "An Experiment in Welfare Capitalism: The Health Care Services of the Tennessee Coal, Iron and Railroad Company." PhD diss., University of Alabama, 1983.

Rogers, William Warren, Robert David Ward, Leah Rawls Atkins, and Wayne Flynt. *Alabama: The History of a Deep South State.* Tuscaloosa: University of Alabama Press, 1994.

Russo, Mary Sicard Cobb. "History of Sicard and Mas Families, as of July 1991." Transcript in the possession of the author.

Sachs, Patricia. "Together We Work, Together We Grow Old: Life, Work and Community in a Coal Mining Town." PhD diss., The City University of New York, 1986.

Scott, Shaunna L. *Two Sides to Everything: The Cultural Construction of Class Consciousness in Harlan County, Kentucky.* New York: State University of New York Press, 1995.

Shannon, Lewis. *Billy Gould: Coal Mine and Coke Oven Site.* National Park Service, Historic American Engineering Record, Birmingham Recording Project, 1992.

Shifflett, Crandall A. *Coal Towns: Life, Work, and Culture in Company Towns of Southern Appalachia, 1880–1960.* Knoxville: University of Tennessee Press, 1991.

Shotts, Reynold Q. *Coal Reserves of Bibb County, Alabama. Geological Survey of Alabama.* Circular 72. University: University of Alabama, Division of Economic Geology, 1971.

———. *A Summary of the Composition and Technology of the Coals of Bibb County, Alabama. Geological Survey of Alabama.* Circular 73. University: University of Alabama,Division of Economic Geology, 1971.

Smith, Eugene A. *Geological Survey of Alabama.* Montgomery, AL: The Brown Printing Co., 1890.

———. *Geological Survey of Alabama: Report of Progress for 1875.* Montgomery, AL: W. W. Screws, 1876.

Squire, Joseph. "Autobiography, 1908(?)." Transcript in the possession of the author.

———. "Diary of Joseph Squire for year 1861." Transcript in the possession of the author.

———. "Joseph Squire's Diary—1868." Transcript in the possession of the author.

———. "Papers and Correspondence." Archives, Birmingham Public Library, Birmingham, AL.

———. "Papers and Correspondence." Private Collection, Ken Penhale, Helena, AL.

State of Alabama. *Annual Report of Coal Mines.* Birmingham, AL: Dispatch Printing Co., 1913–1929, 1934–1938, FY 1944–1948.

State of Alabama, Department of Industrial Relations, Division of Safety and Inspection. *Annual Statistical Report.* Wetumpka, AL: Wetumpka Printing Co., FY 1951–1991.

Stokes, Melvyn, and Rick Halpern, eds. *Race and Class in the American South Since 1890.* Oxford: Berg Publishers, 1994.

Straw, Richard A. "The Collapse of Biracial Unionism: The Alabama Coal Strike of 1908." *Alabama Historical Quarterly* 37, no. 2 (Summer 1975), 92–114.

———. "Soldiers and Miners in a Strike Zone: Birmingham, 1908." *Alabama Review* 38, no. 4 (October 1985), 289–308.

———. "The United Mine Workers of America and the 1920 Coal Strike in Alabama." *Alabama Review* 28, no. 2 (April 1975), 104–128.

Taft, Philip. *Organizing Dixie: Alabama Workers in the Industrial Era.* Westport, CT: Greenwood Press, 1981.

Tarrant, S. F. H. *Hon. Daniel Pratt: A Biography.* Richmond, VA: Whittet & Shepperson, 1904.

Teague, Rev. E. B. "Historical Sketches of Shelby County." Transcript in the possession of the author.

Thornton, J. Mills, III. *Politics and Power in a Slave Society: Alabama, 1800–1860.* Baton Rouge: Louisiana State University Press, 1978.

Tindall, George Brown, and David E. Shi. *America: A Narrative History.* 8th ed. New York: W. W. Norton, 2009.

Troost, Gerard. "The Beginning of Geological Exploration in Northern Alabama: Gerard Troost's Travels in 1834." Edited by James X. Corgan and Michael A. Gibson. *Alabama Review* 48, no. 2 (April 1995), 142–153.

Tullos, Allen. *Habits of Industry: White Culture and the Transformation of the Carolina*

Piedmont. Fred W. Morrison Series in Southern Studies. Chapel Hill: University of North Carolina Press, 1989.

U.S. Department of the Interior. *A Medical Survey of the Bituminous Coal Industry: Report of the Coal Mines Administration.* Washington, DC: Government Printing Office, 1947.

Walker, James H. *The Struggle and the Joy: An American Coal Town—Piper, Alabama.* Birmingham: Birmingham Public Library Press, 1993.

Wallace, Anthony F. C. *St. Clair: A Nineteenth-Century Coal Town's Experience with a Disaster-Prone Industry.* New York: Alfred A. Knopf, 1987.

Ward, Robert David, and William Warren Rogers. *Convicts, Coal, and the Banner Mine Tragedy.* Tuscaloosa: University of Alabama Press, 1987.

———. *Labor Revolt in Alabama: The Great Strike of 1894.* University: University of Alabama Press, 1965.

Webb, Samuel L., and Margaret E. Armbrester. *Alabama Governors: A Political History of the State.* Tuscaloosa: University of Alabama Press, 2001.

Wiener, Jonathan M. *Social Origins of the New South: Alabama, 1860–1885.* Baton Rouge: Louisiana State University Press, 1978.

Williams, Michael E., Sr. *Isaac Taylor Tichenor: The Creation of the Baptist New South.* Tuscaloosa: University of Alabama Press, 2005.

Worthman, Paul B. "Black Workers and Labor Unions in Birmingham, Alabama, 1897–1904." *Labor History* 10, no. 3 (Summer 1969), 375–407.

Wright, Gavin. *Old South, New South: Revolutions in the Southern Economy Since the Civil War.* New York: Basic Books, 1986.

Yearley, C. K., Jr. *Enterprise and Anthracite: Economics and Democracy in Schuylkill County, 1820–1875.* Baltimore: The Johns Hopkins Press, 1961.

INDEX